AVIATION SCHOLARSHIPS !

The Ultimate Resource on Financial Assistance for College, Flight Training, and Career Advancement

By Sedgwick D. Hines

Flight Time Publishing, Chicago, Illinois

Published by:

Flight Time Publishing
8526 Drexel Ave., Suite 3B
Chicago, IL 60619-6210 USA

Library of Congress Cataloging in Publication Data

Hines, Sedgwick D.
Aviation scholarships : the ultimate resource on financial assistance for college, flight training, and career advancement / Sedgwick D. Hines. -- 3rd ed.
p. cm.
Includes index.
LCCN: 99-94720
ISBN: 0-9657384-3-4

1. Aeronautics--Scholarships, fellowships, etc. --United States--Directories. I. Title.

TL537.H56 1999 629.13'0079'73
QBI99-309

Library of Congress Catalog Card Number: 99-94720

ISBN 0-9657384-3-4

Message from the Author

As an Alumni of Purdue University at West Lafayette, IN, Certified Flight & Instrument Instructor, Author and Airline Pilot, I understand the financial hardships students encounter while attending college and flight training. I received over $90,000 in the form of scholarships, grants, and fee remissions (this amount does not include college loans) to help finance my education. I have taken my personal experiences and the help from others to present to you an invaluable tool which should be used during your scholarship search.

One of the biggest difficulties facing college-bound students is struggling to find the money to pay the increasing costs of college tuition, flight training fees, books, supplies, room, and board. By using the information contained in **AVIATION SCHOLARSHIPS**, you'll discover opportunities to utilize private & corporate scholarships, grants, fellowships, and loans to help finance your education.

My advice to anyone interested in obtaining financial aid is to apply to as many scholarships as possible. With **AVIATION SCHOLARSHIPS**, I've done some of the homework for you. This scholarship guide provides you with the information you need to find those valuable, often-overlooked scholarships. This guide gives you inside tips on how to apply and promote yourself as a worthy applicant. You are also provided with the contact names, addresses, phone numbers, eligibility requirements, and much more.

Your comments are appreciated and are taken seriously. Readers are the most constructive critics. Your feedback will help me to improve future editions of this book.

College and flying can be an exciting time, full of personal growth and discovery. I assure you that **AVIATION SCHOLARSHIP** can help you plan to realize your dream of a career in aviation, aerospace, or space exploration.

Acknowledgments

The author and publisher would like to thank the following people for their contributions to **AVIATION SCHOLARSHIP**. Their dedication has helped to make this publication accurate and up-to-date. Awards are subject to change without notice.

I must first give thanks to God who has guided me throughout the completion of this project and to whom I have prayed will bless all who use this guide with success. It is imperative to have a supportive family and group of friends to make any book possible. It is those people in particular who have been optimistic supporters from the being whom I appreciate the most.

I would like to give special thanks to:

Yolanda Hines: Mom, Thank You for your support, dedication and love. Your support was and continues to be the key factor in my childhood dream of becoming a professional airline pilot. Your inspiration has allowed me to achieve my dream and goals. I Love You.

Charmaine and Greg Warren: Having a supporting cast is very important. Your hard work and assistance made this book possible. I truly appreciate your dedication.

The Gilliam Family: I would like to thank your family for the support you'll have given me

Nia Gilliam: Thanks for all of your help, it was fun working together. Remember, there will be other projects.

I would also like to thank The Roberts Family, The Williams Family, Reggie Dunlop, Emily Carter, Eboni Haynes, Mr. and Mrs. Norwood, Leon Johnson, The Organization of Black Airline Pilots, and magazine editors.

I sincerely thank the officials and administrators at the many institutions, foundations, organizations, corporations, and government agencies for their cooperation in completing the request forms. I sincerely thank all these fine people.

THIS PAGE INTENTIONALLY LEFT BLANK

Table of Contents

Disclaimer

This book is designed to provide information in regards to the subject matter covered. It is sold with the understanding that the publisher and author are not engaged in rendering legal, accounting or other professional services. If legal or other expert assistance is required, the services of a competent professional should be sought

The purpose of this guide is to assist students in finding financial assistance for higher education primarily in the United States. Every effort has been made to make this manual as complete and as accurate as possible. However, there may be mistakes both typographical and in content. Furthermore, the author and publisher have made every effort possible to ensure that the listings in this directory are correct at the time of publication. Scholarship funding, guidelines, and application procedures change frequently, as do address and phone numbers. This book should be used only as a research aid in locating scholarships, grants, fellowships, internships, and loan funding sources. It does not guarantee that you will receive any type of aid. The author nor publisher assume liability nor responsibility for any problems that may be encountered due to inaccurate information or to any person or entity with respect to any loss or damage caused or alleged to be caused directly or indirectly by the information contained herein.

THIS PAGE INTENTIONALLY LEFT BLANK

INTRODUCTION

This book is an in-depth reference guide that has valuable information about financial aid opportunities. It is designed to assist and possibly match you with unclaimed aviation scholarships, grants, fellowships, internships, and loans awarded to high school and college students each year. The challenge of paying for college requires planning, creativity, and resourcefulness. If you invest the time required, you'll find there are many ways to manage college costs and many channels through which you can receive help.

This guide is structured to expedite your search for financial assistance. It is divided into several sections covering various aviation related fields and has a personal workbook. Provided below is a description of each section.

Financial Aid Information section provides you with an explanation of award types, federal programs offered to students, and several telephone numbers offered from the U.S. Department of Education, where you can obtain more information about federal student financial aid.

Scholarship Search Process section provides you with various tips on: searching for financial assistance, how to apply for scholarships, requesting scholarship information, completing the application, interviewing tips, sample essay questions, improving your personal skills and qualifications. There is an example of how to keep track of your scholarship process. After you have completed your application, there is a scholarship application checklist that allows you to check off the appropriate items requested by the sponsor(s) for each individual scholarship.

Scholarship Opportunities section is arranged alphabetically by scholarship name followed by the sponsors. Within this section, scholarships are categorized into various fields to expedite and narrow your search for appropriate awards. Each source contains a brief description of the award, often providing users with information they need to make a decision about applying. For further information about the requirements and a more detailed description, write to the contact person(s) and addresses provided. Provided below is a list of categories covered:

- ✈ All Majors
- ✈ Aeronautics & Astronautics
- ✈ Aerospace Science & Technology
- ✈ Administration / Management
- ✈ Aviation Maintenance
- ✈ Avionics / Aircraft Electronics
- ✈ Fellowships
- ✈ Flight Training
- ✈ Grants For Educators And Chapters
- ✈ Military Affiliation
- ✈ School Sponsored
- ✈ Special Interest & Affiliation
- ✈ Women
- ✈ Loans

Internships & Cooperative Education Programs section provides a brief description of how to write a resume, coverletter, and interviewing tips. There is a listing of industry companies and other departments that offer students the opportunity to broaden their horizons by working for the company. Some of the intern positions are salary based, others are not.

FAA Aviation Education Representatives section provides you with a listing of education representatives by regions throughout the United States.

State Aviation & Airport Departments section provides you with a listing of state aviation and airport departments. These departments may offer airport-related scholarships, grants, internships, cooperative education programs, or loans to students. Awards or internships are usually in the aviation safety/education bureaus or in the airport engineering/planning bureau. Some of the internships and cooperative education programs may or may not offer compensation. These state departments may be restricted to their legal residents. However, they may also be available to out-of-state students who may be or are attending public or private colleges or universities within the state. These departments are aware of many opportunities for career advancements.

Space Grant Consortium Directors & Departments section provides you with a listing of Space Grant Consortium Directors and Departments. The goal of the Space Grant Consortium programs is to encourage interdisciplinary training and research, to train professionals for careers in aerospace science, technology, and allied fields, and to encourage individuals from underrepresented groups to consider careers in aerospace fields. These state departments may be restricted to their legal residents. However, they may also be available to out-of-state students who may be or are attending public or private colleges or universities within the state. These departments are aware of many opportunities for career advancements.

State Financial Assistance Agencies section provides you with a listing of state agencies that offers assistance to qualified students. These state agencies may be restricted to their legal residents. However, they may also be available to out-of-state students who will be or are attending public or private colleges or universities within the state. After each state's name is the agency that may provide you with some assistance.

Glossary section presents common definitions for many of the words used by financial aid administrators and scholarship sponsors for students and parents.

Appendix section offers various forms to complete to assist students with their pursuit of receiving financial assistance via government programs, organizations, foundations, individuals, institutions, and companies. The forms are designed to help you expedite the application process by uncovering your skills & abilities, providing sample aviation essay questions, and Scholarship Track Form to chart your progress. The Scholarship Track Form allows you to keep an accurate track of what sources you have applied for, when, who the contact person was, the deadline date, and the results of the inquiry.

Consult the "User's Guide" following this introduction for details on the content, arrangement, and indexing of entries.

USER'S GUIDE

AVIATION SCHOLARSHIPS is comprised of the scholarship opportunities section that contains a descriptive listing on awards and their sponsors. Entries are arranged alphabetically by scholarship name followed by the sponsor. Sources are referenced through several categories:

- All Majors
- Aeronautics & Astronautics
- Aerospace Science & Technology
- Administration / Management
- Aviation Maintenance
- Avionics / Aircraft Electronics
- Fellowships
- Flight Training
- Grants For Educators And Chapters
- Military Affiliation
- School Sponsored
- Special Interest & Affiliation
- Women
- Loans

Each scholarship described in this book are listed alphabetically. School-specific scholarships are listed under the institution's name through which the scholarship is provided.

The description of each scholarship in this book follows a certain format which is divided into sections. The sections gives you important information at a glance. Some scholarships may be described in more detail than others. You are provided the necessary information to contact the scholarship sponsor via letter(s), or telephone calls for additional information.

Please Note

Students are strongly encouraged to read the descriptions carefully and pay particular attention to the various eligibility requirements before applying for awards. Apply only if you meet the eligibility requirements. Do not apply if you do not meet all of the eligibility requirements. If you are eligible, you should enclose a stamped, self-addressed #10 business size envelope when requesting details and applications.

Some scholarship sponsors will not allow us to list their awards, because some applicants do not follow or read the requirements carefully. Sponsors receive many requests from applicants who do not qualify. This creates unnecessary work and expense for sponsors who must answer requests.

When applying for financial assistance, do not send photocopies or computer-generated letters to every source you find. Read each scholarship information carefully. Write or type an individual letter, **ONLY** if you meet the eligibility requirements. The letter should clearly define your qualifications.

The scholarship and award entry shown below illustrates the kind of information that is typically included in this book.

Sample Entry

❒ **David Washington Scholarship** *(Scholarship Name)*
Flight Time Services *(Sponsor)*
4567 W. Ohio *(Contact Address)*
Chicago, IL 60631
Tel. (773) 898-9856
Fax (773) 896-9685

Deadline: April 10

Purpose: To assist aviation students in pursuing a career in the aviation industry.

Eligibility: Preference is given to candidates who are pursuing a career as a professional pilot.

Award(s) & Amount(s): 2, $2,000/award

The data on each scholarship is divided into the following sections:

Checkbox: This maybe used to check-off all the possible awards you may be eligible to apply.

Name: The name of the scholarship..

Sponsor: The name of the institution, organization, company , or individual sponsoring the award.

Contact Person/Department: Some listings, the name of the person or department in charge of providing scholarship information is available. Ask to speak to the contact person when calling. Contact person(s) may change, if this is the case; ask to speak with someone in charge of the scholarship.

Contact Address: Some listings, the address and/or telephone number of the organization, company, etc. that offers the scholarship is provided. This information maybe used to obtain further information about the scholarship and/or to start the application process. Contact information for the school-specific scholarships often appears under the name of the school rather than under the name of each scholarship.

Deadline : This is the final deadline date to send in the application. Some deadlines dates varies from year to year. You are given the month in which the application is due. Contact the sponsor for the exact date to return the application.

Purpose/Description: Some listings provide a purpose or description is the scholarship offered.

Eligibility: : Each scholarship has some basic requirements that applicants must fulfill. **Make sure you read the descriptions carefully and pay particular attention to the various eligibility requirements <u>before</u> applying for an award.**

Notes: This section may give applicants additional information about the scholarship. Make sure you review the notes section.

Award(s) & Amount(s): This section indicates the number of awards given each year. If there is not an award given, contact the sponsor for more details. The amount of money awarded to recipients of the scholarship may vary from year-to-year. Once again, contact the sponsor fro more information.

Compiling Information

Scholarship awards were collected throughout the 1998 year by mailing detailed questionnaires to colleges, universities, foundations, organizations, clubs, and corporations in the United States. Our search also included leads from publications, newsletters, and Internet sites.

We contacted the various sponsors by mailing several questionnaires and by telephone calls. Each questionnaire requested contact information and a brief description of their award(s). The questionnaires that were completed and returned were entered into the appropriate categories. We believe that the awards presented in this book are accurate. Some sponsors failed to respond to our questionnaires before the deadline date. Their award(s) are not included in this edition of the book. Students should always request more information from the sponsors in which they are interested.

Updating This Book

This edition, covering 1999 - 2000, is the 3rd edition of **AVIATION SCHOLARSHIPS.** The next biennial edition will cover the years 2000 - 2001 and will be issued in early 2000.

THIS PAGE INTENTIONALLY LEFT BLANK

Financial Aid Information

This section is designed to assist you and your parent(s) with understanding and receiving financial aid. The following topics will be discussed in detail:

- ✈ Applying for Financial Aid Based on Need
- ✈ How to fill out the financial aid forms
- ✈ Explanation of the most commonly used terms
- ✈ Explanation of Contribution terms
- ✈ Deciding if you qualify for financial aid based on need
- ✈ How to use the *Cost of College Form* and *Financial aid Comparison Form*
- ✈ What is a Financial Aid Award Letter
- ✈ What to do after receiving the award letter
- ✈ Collegiate Awards and Programs
- ✈ Explanation of Federal Aid Awards
- ✈ U.S.A. Federal Aid Special Programs

Financial Aid Information

Applying For Financial Aid Based On Need
Financial aid is given to students who have financial need. A student's financial need is the difference between what you and your parents can pay (Expect Family Contribution) and what it costs to attend a particular school.

The Free Application for Federal Student Aid (FAFSA) is used by the government to determine eligibility to apply for aid based on need. Completing a financial aid application is an easy process. The forms are very similar to income tax forms and require much of the same information. Almost every state that offers student assistance uses the federal government's system to award aid to prospective and current students. The FAFSA is the key to receiving your possible share of the billions of dollars awarded annually to needy students.

A Free Application for Federal Student Aid (FAFSA) can be obtained from your high school guidance counselors, library, or financial aid office. The FAFSA should be available after mid-November of each year. Regardless of your situation, you should always complete and mail the FAFSA for each school year. The form will ask for the current year's financial data. The FAFSA should be filed after January 1, of the new year, in time to meet the earliest college or state scholarship deadlines. These forms also contain a place to apply for a Federal Pell Grant.

The College Scholarship Service or American College Testing Program takes the information on the FAFSA and uses computers to estimate your family's expected contribution. They send this estimate to the colleges you listed on your application.

Within two to four weeks after you have mailed the form, you will receive a summary of the FAFSA information called the Student Aid Report (SAR). The SAR will give you an estimated family contribution and also allow you to make corrections (if needed) to the data you submitted. The schools you listed on the FAFSA will also receive this information. Some colleges will also expect you or your parents to fill out additional financial aid application forms. Be sure to check with the financial aid office.

Each college will put together a financial aid package for you. Each one will calculate how much money you will need to attend that school and will decide where the funding will come from. A financial aid package typically includes aid from at least one of the major sources: federal, state, college, loans, and family contribution.

You will not receive a financial aid package from a school until after you have been admitted. You could be admitted to a school and not get financial aid. Always try anyway. Apply for financial aid, and then wait.

How To Fill Out The Financial Aid Form (FAF)
Read the financial aid information you received from each school you have applied to for admission.

You should ask yourself the following questions:

- Does the college require the FAF? Get these forms from your counselor. (You can also check the information that comes with the FAF. They list the schools that use their form.)
- Does the college have a separate financial aid form for you to fill out?
- Do you have any applications from outside scholarship sponsors you want to apply for?

Deadlines!
Write the dates on the calendar. Allow yourself enough time to finish the applications and extra time to mail them.

Before you start filling out the financial aid forms, collect all your family's financial documents. You should have:

- Your parents' most recent United States income tax return, and your own. W—2 form(s) and any other documents of money earned during the last year.
- Current bank statements.
- Current mortgage statements (if your family owns a house or an apartment).
- Records of medical and dental bills for the last year.
- Records of veteran's benefits and Social Security payments.

Your parents may choose to fill out the FAF themselves or you may do it together. Your parents must sign the completed forms, if you are a dependent student.

Filling out financial aid applications can be confusing. Just fill out one form at a time and answer one question at a time. If you really get stuck, write to the college Financial Aid Office or ask your counselor for help.

When you beginning the application process, there are several steps you and your parents should do.

- Make a practice worksheet. Either copy the form itself or use blank paper. Answer every question before you write anything on the form itself. Use pencil on your practice form so you can change your answers easily.
- Read all the instructions twice before you start.
- Answer every question. Follow the instructions on what to do if a question does not apply to you.
- Be neat. (Mistakes can happen, if your answers cannot be read.)
- Give accurate information. Be honest. Your financial aid application may be checked against your parents' income tax return.

Be sure to show any ***unusual financial situations***. These include:

- Medical expenses (because of a serious or long illness)
- A handicapped member of the family
- Unemployment
- Separation or divorce
- A parent who is disabled.
- A parent who is retired.
- A parent who has died.
- More than one child in college at the same time, or a parent who is a full time student.
- Debt (but not consumer debt for things you've bought)

If you are applying for a Federal Pell Grant, be sure to check the box on the FAF. You do not pay a fee to apply for a federal grant.

List the colleges you want to receive the FAF report. There is a fee for sending a report to each college you have listed.

When you have completed your practice worksheet, copy the answers neatly and carefully on the original form.

Make sure your parent(s) have signed the form in the proper place(s).

- Include a check / money order for the correct amount of reports you want sent to the colleges listed.
- Make a copy of the original form for your files.
- Mail the original
- Make sure the return address is accurate.

Some FAF's will be validated. You will be asked to send a copy of your family's income tax return, so the accuracy of the information can be checked. You must send the tax return if you are asked to do so, or you will NOT receive financial aid. Do not send any income tax forms, unless they are requested.

If your financial situation change, write to the Financial Aid Office of every college you have applied and to the offices of each private scholarship organization. Imform them of exact amount (in dollars and cents) of how your situation has changed.

After you have applied for admission and for financial aid, each college will notify you about whether you have been admitted and about whether you will receive financial aid. Read these letters carefully.

Questions to Ask Yourself After Reviewing Your Financial Aid Package or Acceptance Letter for Admission

- When do you have to let the college know if you will attend?
- Is a deposit required?
- Is it refundable? Until when?
- When do you have to let them know if you want to accept the financial aid package?
- Is a housing deposit for a dormitory room required?

If you are accepted by more than one school, you will have to decide which one to attend. This can be a hard decision.

THINK ABOUT WHICH SCHOOL WILL BE BEST FOR YOUR EDUCATION AND YOUR CAREER PLANS.

Explanation Of The Most Commonly Used Financial Aid Terms
There are several terms you and your parent(s) need to become very familiar with. Provided below is a brief explanation of some common terms.

COLLEGE FEES

- *Tuition* - the cost of registering for college courses. Tuition is called "General Fees" at some colleges. State colleges charge two different rates for tuition. In—state tuition is a lower tuition fee for students who live in that state. Out—of—state tuition is the higher fee paid by students who live in a different state.

- *Room and Board* - charges are for a dormitory room and meals. Some schools offer a choice of meal plans. Are you paying for the meal plan you want? Select the appropriate meal or "board" plan of your choice. At some schools you must pay for three meals a day/seven days a week. Some schools offer a choice of meal plans you may be able to select (i.e. two meals a day/five days a week). If you have a choice, think about your schedule and your budget.

- *Student Activities Fees* - is used to support student activities such as athletics, student newspaper, movies, student government, etc. Sometimes this fee is included in the tuition or general fees.

- *Health Fees* - pay for medical care at the college hospital. Some schools bill students for each visit to the hospital or clinic rather than charging a health fee.

- *Laboratory Fees* - may be charged for courses such as flight training, simulator training, ground school, physics, or chemistry that require use of a laboratory.

- *Miscellaneous Fees* - may be charged depending on the courses you take. For example, there may be a studio fee for art courses, or a fee for music lessons or instrument rental(s).

CHECK THE FEES ON YOUR BILL

If you are charged lab or miscellaneous fees, make sure the fees are for courses you are taking. Check your bill for any mistakes.

Explanation of Contributions Terms

Family contribution is the amount of money a family will be expected to contribute toward college costs for their dependent student. (A dependent student is one who depends upon his/her parents for support. Both the parents' and student's income and assets are evaluated to determine the family contribution to college costs.)

- The contribution from parental income includes taxable and nontaxable income.
- The contribution from parental assets includes stocks, bonds, savings, and business assets.
- The contribution from student income includes taxable and nontaxable income.

Parent Contribution - This will be based on your family's financial situation. The information provided in the Financial Aid Form (FAF) which your parents must fill out determines what they will be expected to pay. Your family's expected contribution is based on family income, the number of children in your family, the cost of living, assets, certain debts and special circumstances.

Student Contribution - Your contribution will come from: Jobs (summer, part-time, or weekend jobs), Savings accounts and other outside income that you may expect to receive to contribute to your college expenses

Deciding Whether You Qualify For Financial Aid, Based On Need

You should begin to think about how much college will cost and where the money will come from. If you do not know the exact amount, "Guesses" are okay. The college catalogue will give you information about expenses at that school. The *Cost of College Form* and *Financial Aid Comparison Form* maybe used to provide you with an ***estimate*** only. The completion of each form does not guaranteed the actual college tuition or Financial Aid Package will be the same as your estimates.

Step 1: You need to evaluate whether or not you (the student) qualify for financial aid based on need. This is a critical step. An estimated $40 billion in student aid is awarded to needy families. To judge your chances of receiving aid based on need, estimate your expected family contribution, then use the *Cost of College Form* and *Financial Aid Comparison Form* provided on the next page to evaluate your situation.

Cost of College Form

	College 1	College 2	College 3
Name of college	________	________	________
Tuition	________	________	________
Fees:	________	________	________
- Student activities	________	________	________
- Medical services	________	________	________
- Laboratory	________	________	________
- Other	________	________	________
Room & Board	________	________	________
- On Campus	________	________	________
- Home	________	________	________
Books & Supplies	________	________	________
Personal Expenses	________	________	________
Travel	________	________	________
- Campus	________	________	________
- Holidays	________	________	________
Total Budget	+ ________	+ ________	+ ________

Step 2: Fill out this form *after* you have received your financial aid package. When you think about which financial aid package to accept, you must compare how much is a grant, how much is a loan and how much is money from work. The largest package may not be the best one for you. *Carefully* look at the difference between Total A (expenses), and Total B (family contribution and grants). The amount remaining is what you must borrow or earn.

Financial Aid Comparison Form

Name of College: ____________________________

EXPENSES:

Tuition and Fees $ ______________

Room and Board $ ______________

Books and Supplies $ ______________

Transportation $ ______________

Clothing $ ______________

Miscellaneous $ ______________

TOTAL A $ ______________

FINANCIAL RESOURCES

Student Contribution $ ______________

Parents' Contribution $ ______________

Supplementary Education Opportunity Grant $ ______________

Pell Grant $ ______________

Other Grants: $ ______________

Private Scholarships

________________ Scholarships $ ______________

________________ Scholarships $ ______________

________________ Scholarships $ ______________

TOTAL B $ ______________

LOANS

Stafford Loan (formerly GSL) $ ______________

Perkins Loan (formerly NDSL) $ ______________

Other Loans $ ______________

College Work Study Program $ ______________

TOTAL C $ ______________

Money you must borrow or earn: Total A - Total B = $ ______________

Total Financial Aid Package: Total B + Total C = $ ______________

Unmet Financial Need: Total A - (Total B + Total C) = $ ______________

What Is A Financial Aid Award Letter
This letter is issued by the Financial Aid Office that lists all of the financial aid awarded to the student for the academic year. This letter provides details of your financial aid package according to amount, source, and type of aid. The award letter will include the terms and conditions for the financial aid and information about the cost of attendance. You are required to sign a copy of the letter, indicating whether you accept or decline each source of aid, and return it to the financial aid office. Before you mail it, make sure you make a copy for your files.

What To Do After Receiving The Financial Aid Award Letter
Since many colleges award financial aid to students on the basis of first come first served, mail your application as soon as possible to increase your chances of receiving the maximum aid for which you are eligible.

You should respond promptly to the award letter, making sure to enclose all the requested information. Failure to respond by the deadline may result in denial or reduced financial aid.

Receiving Aid Not Based On Need
Whether or not you qualify for a Need-Based award, it is always worthwhile to look into Merit scholarships from outside sources such as foundations, agencies, religious groups, and service organizations. If a student that is not eligible for Need-Based award, the student should apply for a Merit scholarships. Only apply for this scholarship if you are eligible. If you later qualify for a Need-Based award, a Merit scholarship provides benefit of reducing the loan and/or work study portion of an award.

Most Merit scholarships are highly competitive. Use the following guidelines when investigating Merit scholarships.

- Take advantage of any scholarships for which you are eligible based on employer benefits, military service, church membership, other affiliations, and student or parent attributes (ethnic background, nationality, etc.). Company or union tuition remissions are the most common examples of these awards.
- Scholarship directories are useful resources that can be found in the high school or public library. Computerized scholarship searches provide essentially the same service but can cost anywhere from $10 to $100.
- Investigate community scholarships

In addition to conducting a search for Merit scholarships, there are loans and job opportunities for students who do not qualify for aid based on need.

Collegiate Awards And Programs
Most colleges, universities, and other educational institutions offer their own financial aid programs and payment plans. Their financial aid offices may have information on privately sponsored awards that are specifically designated for students attending those institutions. Contact the financial aid offices of all the institutions in which you have an interest. Request an application(s) and detailed information on all of the financial aid programs that they sponsor or administer in your desired field of study. Inquire about special awards (such as non-traditional scholarships, minority or ethnic awards, fee remissions, academic recognition, prepayment plans, discounts for alumni children and student leaders, recruiting discounts, and tuition equalization).

Explanation Of Federal Aid Awards
Federal aid for college students is available through a variety of programs administered by the U.S. Department of Education. Most colleges and universities participate with federal programs, but there are exceptions. Contact the school's financial aid office to find out if it is a participating institution. If the school is a participant, the student should work with the financial aid counselors to determine how much aid can be obtained.

Financial aid comes in three forms: grants (gifts to the student), loans (funds which must be repaid), and work-study jobs (employment during enrollment in school). Provided below are various types of grants, and loans.

Types of Grants
- Pell Grants
- Federal Supplemental Educational Opportunity Grants (FSEOG)

Types of Loans
- Direct Loan Program
- Federal Family Education Loans (FFEL) / Stafford Loans
- Direct and FFEL / Stafford Programs Loans for Parents (PLUS)
- Perkins Loan Program

Explanation Of Federal Aid Awards

FEDERAL PELL GRANT
The Federal Pell Grant is the largest grant program; over 5 million students receive awards annually. This grant is intended to be the base, or starting point, of assistance for lower-income families. Eligibility for a Federal Pell Grant depends on the Expected Family Contribution (EFC). The amount received will depend on the family's contribution, the cost of education at the college or university, and whether the student will attend full-time or part-time. The highest award depends on how much the program is funded. Check the box on the FAF form to apply for this money. This is the biggest federal program to help students.

FEDERAL SUPPLEMENTAL EDUCATIONAL OPPORTUNITY GRANT (FSEOG)
The Federal Supplemental Educational Opportunity Grant (FSEOG) provides additional need based federal grant money to supplement the Federal Pell Grant. Each participating college is given funds to be awarded to needy students. The maximum award is $4,000 per year, but the amount received will depend on the college's policy, the availability of FSEOG funds, the total cost of education, and the amount of other aid awarded.

FEDERAL WORK-STUDY (FWS)
This program provides jobs for students who need financial aid for their educational expenses. The salary is paid by funds from the federal government and the college (or the employer). The student works on an hourly basis and must be paid at least the federal minimum wage. The student may earn only up to the amount awarded, which depends on the calculated financial need and the total amount of money available to the college.

FEDERAL PERKINS LOAN
This loan is a low-interest (5 percent) loan for students with exceptional financial need (students with the lowest Expected Family Contribution). Federal Perkins Loans are made through the college's financial aid office. Students may borrow a maximum of $3000 per year for up to five years of undergraduate study. They may take up to ten years to pay the loan, beginning nine months after they graduate, leave school, or drop below half-time status. Interest on this loan does not accrue while the student is in school, and under certain conditions (e.g. teach in low-income areas, work in law enforcement, full-time nurses or medical technicians, serve as a Peace Corps or VISTA volunteers, etc.). Some or all of the loan can be canceled or payments deferred.

FEDERAL STAFFORD LOAN
A Federal Stafford Loan may be borrowed from a participating commercial lender such as a bank, credit union, or savings and loan association. The interest rates vary annually (up to a maximum of 8.25 percent), and the rate for 1998-99 is 7.66 percent. If the student qualifies for a need-based subsidized Federal Stafford Loan, the interest is paid by the federal government while the student is enrolled in school. There is also an unsubsidized Federal Stafford Loan that is not based on need.

The maximum amount dependent students may borrow in any one year is $2625 for freshmen, $3500 for sophomores, and $5500 for juniors and seniors, with a maximum of $23,000 for the total undergraduate program. The maximum amount independent students can borrow is $6625 for freshmen (no more than $2625 in subsidized Stafford Loans), $7500 for sophomores (no more than $3500 in subsidized Stafford Loans), and $10,500 for juniors and seniors (no more than $5500 in subsidized Stafford Loans). Borrowers must pay a 4 percent fee, which is deducted from the loan proceeds.

To apply for a Federal Stafford Loan, the student must first complete a FAFSA to determine eligibility for a subsidized loan, then a separate loan application is submitted to a lender. The financial aid office can help the selection process of a lender or you can contact your state department of higher education to find a participating lender. The lender will send a promissory note that the student must sign agreeing to repay the loan. The proceeds of the loan, less the origination fee, will be sent to the student's school to be either credited to the student's account or paid to the borrower directly.

If the student qualifies for a subsidized Federal Stafford Loan, he/she does not have to pay interest while in school. For an unsubsidized Federal Stafford Loan, the student will be responsible for paying the interest from the time the loan is established. However, some lenders will permit borrowers to delay making interest payments and will add this interest to the loan. Once the repayment period starts, borrowers of both subsidized and unsubsidized Federal Stafford Loans will have to pay a combination of interest and principal each month for up to a ten years.

FEDERAL PLUS LOAN

The Federal PLUS loan is a loan for parents of dependent students designed to help families with cash-flow problems. The loans are made by participating lenders. The loan has a variable interest rate that cannot exceed 9 percent (the rate after January 1, 1999 is 8.26 percent). There is not a yearly limit; you can borrow up to the cost of the student's education minus other financial aid received. Repayment begins sixty days after the money is advanced. A 4 percent fee is subtracted from the proceeds. Parent borrowers must generally have a good credit to qualify for Federal PLUS Loans.

The PLUS loan will be processed under either the Direct or Stafford system, depending on the type of loan program for which the college has contracted.

FEDERAL DIRECT STUDENT LOANS

The Federal Direct Student Loan is a relatively new program which is similar to the Federal Stafford Loan. The primary difference is that the U.S. Department of Education is the lender rather than a bank. Not all colleges participate in this program, and if the student's college does not, he or she can still apply for a Federal Stafford Loan.

Many of the terms of the new Federal Direct Student Loans are similar to those of the Federal Stafford Loan. In particular, the interest rate, loan maximums, deferments, and cancellation benefits are the same. However, under the terms of the Federal Direct Student Loan, students have a choice of repayment plans. They may choose either a fixed monthly payment for ten years; a different fixed monthly payment for ten years, a different fixed monthly payment for twelve to thirty years at a rate that varies with the loan balance; or a variable monthly payment for up to twenty-five years that is based on a percentage of income. Students cannot receive both a Federal Direct Student Loan and a Federal Stafford Loan for the same period of time, but may receive both in different enrollment periods.

For more information about Federal Student Financial Aid, write to:

Federal Student Aid Information Center
P.O. Box 84
Washington, D.C. 20044

Ask for The Student Guide to Financial Aid from the U.S. Department of Education.

Telephone Numbers:
- 1-800-433-3243 to have questions answered about how to apply;
- 1-319-337-5665 to find out if your application has been processed;

• 1-800-730-8913 (TDD) if you are hearing impaired; and
• 1-800-647-8733 to report fraud, waste, or abuse of federal student aid funds.

U.S.A. Federal Aid Special Programs

ROTC
Federal government money has helped thousands get their college education. The Reserve Officers Training Corps in many colleges and universities has assisted numerous students. Current information may be obtained by writing to the individual institutions or ROTC headquarters at:

Army ROTC
Headquarters Cadet Command
Fort Monroe, VA 23651-5000

Navy-Marine Corps ROTC
801 North Randolph St.
Arlington, VA 22203-9933

Air Force ROTC (ATC)
Advisory Service
Maxwell Air Force Base, AL 36112-6663

SERVICE ACADEMIES
Admission to the United States Military Academy at West Point, the United States Naval Academy at Annapolis, and the United States Air Force Academy at Colorado Springs is by appointment and involves competitive exams. There are also appointments made by the President and members of the Congress for certain numbers of enlisted regulars and reservists and a few others.

Admission to the United States Coast Guard Academy at New London and to the United States Merchant Marine Academy at Kings Point is on the basis of competitive examination.

THIS PAGE INTENTIONALLY LEFT BLANK

Scholarship Search Process

Purchasing this book is a great first step in tapping the tremendous scholarship opportunities available nationwide. But you can not stop there. You must take action if you want to see results.

After you have used this book and reviewed some of the financial aid sources to which you would like to apply, you can begin the application process by letting the sponsorship committees know you are interested and qualified for the opportunity. This process includes requesting application materials, completing forms, writing essays and in some cases being personally interviewed. By using this book, your efforts could reap substantial rewards. If you follow the guidelines included in this introductory section, you will minimize the difficulty of applying for aid and increase the confidence with which you present yourself, resulting in more awards and more money for college.

The most important thing to remember when communicating with various organizations is to convey that you are the most qualified applicant. The only things a committee will know about you will be learned through your contact with them. Every call you make or letter you write should demonstrate that you are a capable, professional, and worthy applicant. Scholarship foundations receive large volumes of applications each year, you must make your application stand out.

Each of these subjects will be addressed individually over the next few pages. Apply to as many organizations as time permits and then wait for the positive responses.

- Explanation of Scholarship Award Types
- Tips for Finding Financial Assistance
- Additional Sources to Consider for Financial Assistance
- How to Apply for a Scholarship or Grant
- How to Request Scholarship Information
- Sample request Letter
- The Follow-Up Letter After No Response from the Scholarship Sponsor
- Using the Scholarship Track Form
- Completing the Application Process
- Identifying Your Achievements, Qualifications, and Skills
- Ways to Increase Your Qualifications
- How to Write An Effective Essay
- Actual Essay Questions from Scholarship Applications
- Selecting Someone for a Recommendation Letter
- What to Do After mailing the Application
- Preparing for the Interview
- What to Do After Wining A Scholarship
- Renewing a Scholarship

START YOUR SEARCH EARLY!!

Some awards are given on a first come, first serve basis.

Explanation Of Scholarship Award Types

Every scholarship can be classified as a Local Award, National Award, Merit Award, Need-Based Award, and School-Sponsored Award. A brief description of each award type is provided below. After reading the descriptions, you will have a better understanding of the difference types of awards as you begin to search for scholarship opportunities. Scholarships and grants are awards that do not require repayment.

All of these scholarship types maybe sponsored by companies, public or private organizations, foundations, individuals, and/or colleges and universities.

Local Awards

Local Award amounts varies and are generally given on a one-year basis. Local organizations often sponsor students attending a specific school, state institution, descendants of a local school or organization (students majoring in aviation at the University of North Dakota, descendants of Purdue University Alumni, undergraduates attending a college or university in Southern Illinois, etc.).

Local scholarships may not specify an academic major. Incoming students should consider these scholarships. Applying for a local scholarship is an excellent place to start.

National Awards

National Award amount varies. They generally pay more than Local Awards and attract eligible applicants from all regions. Since eligible applicants are from all parts of the world, National Awards are more competitive than Local Awards. National Awards generally sponsor students considering a specific field of study or profession/career.

Merit Awards

Merit Awards focus on the applicant's achievements. Sponsors are interested in sponsoring students with strong academic records, high grade-point averages, academic and professional goals, student organization involvement, and/or volunteer experience.

Need-based Awards

Need-based awards take into account the financial need based on the information submitted by the applicant on the FAFSA.

Note:

The institution's Financial Aid Office you plan to attend must be informed of outside awards received from private organizations (clubs, union, etc.). Winning a scholarships or grant may change your financial situation, if you are receiving *financial aid based on need.*

School-Sponsored Awards

This section provides you with a listing of colleges, universities, and flight schools who sponsors their own awards. Students are awarded these scholarships, if they attend a specific school and meet the eligibility requirements. If you are interested in attending a specific school, look up the school in this section and there you will find a list of scholarships offered to their prospective and/or current students. This list is not inclusive. If you are interested in a particular school, contact the admissions office or aviation/aeropsace department directly to obtain more information.

Tips For Finding Financial Assistance
There are several strategies to consider for financial assistance. Provided below is a listing of sources to consider for your scholarship search.

- Your parent's union or professional organization.
- A teacher, guidance counselor, or professor may know of a variety of organizations and companies offering scholarships or financial assistance opportunities.
- Consider contacting various civic, fraternal, religious, and business organizations in your community. Many of these organizations have scholarships that are not well publicized. These awards are also usually small, but they add up quickly. These organizations are listed in the yellow pages or on a list at your local Chambers of Commerce.
- Employers represent another major local source of scholarships. If you are working part-time for a company that offers scholarships to its employees, check to see if you can apply for these scholarships. Ask your supervisor to check the company's headquarters for possible opportunities. Inquire about tuition remission scholarships from your employer.
- Your parents' employer may offer scholarships to their employees children. Have your parents check with their benefits/personnel office to see if any scholarship opportunities exist.
- Contact business and industrial groups that provide services or products in your major or career field. This allows you to see if they sponsor students who are entering their profession or industry.
- Contact successful leaders in your community to see if they are sponsoring students or know of associates who sponsor students to go to college.
- Search the major companies, colleges, or universities of your choice on the internet. Many schools have their own web site that provides invaluable information about scholarships programs and much more.

Additional Sources To Consider For Financial Assistance

There are other sources to consider for funding besides the federal or state governments. Consider the following sources: academic departments, faculty members, professors, and state education agencies. Provided below are reasons why you should consider these sources.

- *Academic departments* with which you are affiliated may offer scholarships. Some departments offer their own scholarships in which the financial aid office may not know about.

- *Faculty members* may know about scholarships specifically for your major.

- *Professors* are in contact with outside sources that may offer scholarship information or they may know of another scholarship sponsor to refer you too.

- *FAA Aviation Education Representatives* section (page 141) provides you with a listing of education representatives by regions throughout the United States. These representative may offer or know of some organizations that offers assistance.

- *State Aviation & Airport Departments* section (page 143) provides you with a listing of state aviation and airport departments. These departments may offer airport-related scholarships, grants, internships, cooperative education programs, or loans to students. Awards or internships are usually in the aviation safety/education bureaus or in the airport engineering/planning bureau. Some of the internships and cooperative education programs may or may not offer compensation. These state departments may be restricted to their legal residents. However, they may also be available to out-of-state students who may be or are attending public or private colleges or universities within the state. These departments are aware of many opportunities for career advancements.

✈ *Space Grant Consortium Directors & Departments* section (page 149) provides you with a listing of Space Grant Consortium Directors and Departments. The goal of the Space Grant Consortium programs is to encourage interdisciplinary training and research, to train professionals for careers in aerospace science, technology, and allied fields, and to encourage individuals from underrepresented groups to consider careers in aerospace fields. These state departments may be restricted to their legal residents. However, they may also be available to out-of-state students who may be or are attending public or private colleges or universities within the state. These departments are aware of many opportunities for career advancements. Contact a director or the Consortium office for more information and deadline dates.

✈ *State Financial Assistance Agencies* section (page 157) provides you with a listing of state agencies that offers assistance to qualified students. These state agencies may be restricted to their legal residents. However, they may also be available to out-of-state students who will be or are attending public or private colleges or universities within the state. After each state's name is the agency that may provide you with some assistance.

You can obtain current information from your counselor's office or contact your state's financial assistance agency (see the State Financial Assistance Agencies on page 157) to investigate more opportunities.

What is a Loans?
A loan is money that is borrowed. It must be repaid plus interest. Interest is extra money you pay for using the money you borrowed. Most loans arranged through colleges have low interest rates and do not need to be repaid until you finish college.

There are four things you must know about a loan. You should always ask the following questions.

1. What is the interest rate?
2. Is the interest rate Fixed or Variable?
3. When does the loan have to be repaid?
4. How long can you take to repay it?

Remember:
Networking and developing the right contacts are the best ways to start your scholarship search.

When you are inquiring about any source of assistance (via scholarships, grants, loans), you should:
1. Carefully assess your particular needs and preferences.
2. Consider any special circumstances or conditions that might qualify you for aid geared for a special audience.
3. Carefully research available aid programs.
4. If you have a mentor, ask him/her for information about scholarships or sponsors.
5. Use the telephone. Write letters. Find out what is out there!! **DO NOT WAIT!!!**

Important Note

It is very important to start your scholarship search early. Review all of the scholarships' eligibility requirements before contacting the scholarship sponsors. Create a folder of scholarships you are not eligible for this year. You may be eligible for the same scholarship next year. A very important note to remember - Be aware of the scholarship deadline date. If the deadline date has passed, do not eliminate the possibility of applying for the scholarship. The deadline date may have been extended. Contact the sponsor for more details. Your scholarship search should be continued until you have completed you certificate(s) and/or degree(s).

How To Apply For A Scholarship or Grant
The key to successfully receive financial assistance for school and/or flight training is to apply for as many different awards as possible. Your first step is to write a letter requesting information about each scholarship. Each year, the contact person of a scholarship may change. If a telephone number is available, contact the sponsor before sending your request letter. This will eliminate the processing delay of receiving an application. The key to calling the sponsor is to make sure the request letter goes to the correct person or department. If there is a specific person to address the letter, make sure the person's title is correct (Dr., Mrs., etc.). If there is not a telephone number listed, you should address the letter to the name of the scholarship.

Keep your letter brief and to the point. Remember the purpose of your request letter is to request an application and more details. Your letter should briefly describe how you will use the scholarship money and the school year. Be sure to include your correct contact number in case the sponsor(s) wants to contact you.

Some organizations offer more than one scholarship. If this is the case, you must specify which scholarship(s) you are requesting. If you are requesting multiple applications for different scholarships by the same sponsor, use a separate letter and envelope for each scholarship.

An effective request letter should contain the following elements:

- Date (This allows you to keep track of when the letter was sent.)
- Your name, address, phone number, fax number, e-mail address (as appropriate)
- Scholarship sponsor's name, coordinator (if known), address
- Interest in the scholarship being offered
- Request for application packet
- Expressions of appreciation for a timely response
- "Thank you, " or "Sincerely," to close the letter
- Sign the letter before mailing it
- Include a self-address stamped envelope for the sponsor's convenience in replying.

Remember: TIME IS CRITICAL!!!

Important Tips:

- Have someone proofread your letter for grammatical and spelling errors
- If possible, use a word processor instead of a typewriter

Sample Request Letter

(Date)

(Your name)
(Address)
(Phone Number)

(Scholarship Coordinator - if listed)
(Scholarship name)
(Scholarship Address)

Dear Scholarship Coordinator (use name if known),

I am interested in learning more about your (Scholarship name) program. As a potential applicant, I would greatly appreciate an application or any materials that you could forward to me at this time. I am seeking financial assistance for the (year requesting assistance) academic year. I have enclosed a self-addressed, return envelope for your convenience.

Sincerely,

(Sign Your Name)
(Print Your Name)

* *This is only an example of a request letter.*

The Follow-Up Letter After No Response From The Scholarship Sponsor

Once you have mailed your initial letters, you must keep an accurate track of all your responses and those that have not responded. The *Scholarship Track Form* (located in the appendix section page_) will help you keep an effective track record. Simply follow the instruction on the form.

If you have not received a response after **six** to **eight** weeks, recreate another letter to send again to the scholarship sponsor. It is acceptable to use the original letter, but you must change the date. Do not indicate that you are sending a second letter to the sponsor. You do not want to give the sponsor the wrong impression of you. As a reminder, many of the scholarship sponsors are large organizations that receive many requests; there may have been a chance that your first letter was lost in the paper shuffle.

If you sent a second letter and still have not received a response, do not send another letter. The organization may no longer offer the scholarship. If there is a telephone number available, call the sponsor to inquire about the existence of the scholarship. If there is not a telephone number listed, use directory assistance. Once you call, **remember to be very polite**.

There are several questions you should ask the organization. These questions or suggestions are **only suggestions**; use common sense when speaking with a scholarship sponsor.

- Ask a general question about the scholarship.
  - An example: "I would like to know if your organization offers any scholarship(s) to aviation students?
- Ask about the eligibility requirements and deadline date(s).
- If you are eligible for the scholarship, request an application.
- Ask the question, "How long will it take to receive the application?"

Be sure to use the *Scholarship Track Form* to keep track of your scholarship search progress. Use the notes section to record notes and conversations on the status of your search. Make sure you take note of the person's name, date, and brief conversation. This method should keep your stress level to a minimum.

Scholarship Track Form

The purpose of this form is to assist you with ensuring that you send all the appropriate information requested by the sponsors and to keep track of your progress. Use the checklist(s) for each scholarship before you send off the information.

Use to top section to write the scholarship name, contact person, and very important - deadline date.

Use the Page No. space to easily refer back to the scholarship you're applying too.

The next lines allows you to keep track of the dates you completed each step. Especially when you completed and mailed the application.

This section is to assist you with ensuring that you send all the appropriate information requested. Place an "X" in the appropriate area(s) requested and completed. Place an "NR" in the area(s) that are not requested.

Use the Special Note Section to write down additional dates, times, and contact persons you've been in contact with. Results section should be use to indicate whether or not you won the award.

Scholarship Track Form©

Scholarship Name: ______________________

Contact Person: ______________________

Deadline: ______________________

Page No.: __________ / __________

Request Letter: __________ / __________

Application Received: ______________

Application Completed/Mailed:

__________ / __________

___ Application Complete
___ Official Transcript
___ Letter(s) of Recommendation
___ Essay(s)
___ Reference Sheet
___ Parents Tax Returns
___ Return Envelope
___ Application Signed & Dated
___ Make Copies
___ Other: ______________

Special Notes: ______________________

Results: ______________________

Starting The Application Process
Allow **two** to **four** weeks after mailing your request letters. You should start receiving responses from the various organizations. The application process for individual awards will vary. Allow enough time to complete the application. Pay attention to the application details and deadlines. Review all of the tips provided below:

- ✈ Read all instruction carefully
- ✈ Take note of application deadlines
- ✈ Be very **Neat** and **Professional**
- ✈ Make several copies of the blank application before you start typing on the original.
- ✈ Use the copies as drafts before the final edition. Allow someone to review your draft(s) for mistakes, spelling or grammatical errors. Avoid using correction fluid.
- ✈ Use a word processor or type writer to complete the original application. If you do not have a word processor or type writer. Your local library, school, or friends/parents may have a word processor or type writer at home or office.
- ✈ Do not hand write your application.
- ✈ Accurately and completely enclose all required supporting material (essays, resume, transcripts, recommendation, application). Failure to provide requested information by submitting an incomplete application may result in disqualification.
- ✈ Give references enough time to submit their recommendations.

*Use the *Scholarship Track Form* and Notes Section to keep track of your scholarship inquires and results.

Some sponsors require supplemental information from other sources. Have all the necessary information prepared to successfully complete all the parts of the application. Additional information may be needed such as:

- ✈ Tax Returns (Form 1040) from your parents
- ✈ Academic transcripts (high school, college or both)
- ✈ Official SAT or ACT score(s)
- ✈ Letters of recommendations from academic advisors, teachers, employers, etc.*

An *official academic transcript* is a record of your grades to date at your high school or college/university. When requesting your official transcript, you must specify whether the school must send it directly to the scholarship sponsor or if it should be given to you to be mailed along with your application. Your college/university may charge a fee in order for you to obtain this information. Request several official academic transcripts at the beginning of your scholarship search. Once you have received your transcripts, keep the transcripts in an oversized envelope (i.e. 9" X 12"). This will prevent the transcripts from getting dirty, remember - first impressions are a lasting impression. So be NEAT! Having several transcripts in your possession will help you save time.

Do not get confused. Your official transcript is not the same as your report card obtained by your institution each semester or quarter. As you progress through each semester/quarter, make sure you obtain an up-dated transcript that includes your latest semester/quarter. If you have attended more than one school, you will need to request transcripts from each school. It is important that the scholarship sponsor understand the school's grading scale (i.e. an A is worth 4 points, a B is worth 3, etc.). This will allow them to compare you with other candidates.

Official SAT and ACT scores must be requested from The College Board. The scores should come sealed in an envelope. Order several copies of your scores. This will also save you time. If you retake a test and your scores change, get new copies as soon as possible. Keep the your official score(s) in an oversize envelope.

If *letters of recommendations* are needed, ask people who will provide a positive impact and influence on your application. Some recommendations may need to be tailored to the specific application. Find out if the recommendations are to be sent along with your application or to be sent by the person writing the recommendation. If your recommendation(s) have to be sent separately, inform the reference to wait until you send your application to the sponsor.

Make sure you inform the person (the individual writing your recommendation) when you are going to mail your application to the sponsor. If you have delayed your mailing, be sure to inform the person writing the recommendation. Allowing your application to arrive to the sponsor first will make it easier for the scholarship sponsor to match up all your documents (Academic transcripts, letters of recommendation, etc.). If the sponsor does not have a completed package or documents misplaced, your application may be eliminated. Be sure to read the *Recommendation's Section.*

Don't wait until the last minute to gather this information. Academic transcripts and letters of recommendation may take a few weeks to receive. The importance of **planning ahead** can not be overemphasized.

Important Tips:

✈ Make photocopies of all materials you send out, then create a file for each scholarship.
✈ Make copies of completed applications, use them for future references.

Identifying Your Achievements, Qualifications, Skills, Etc.
This section is designed to help you uncover your skills, abilities, and qualifications which illustrate that you are an ideal candidate for the scholarship. Complete the *Qualifications Form* (appendix section page _) to list or summarize your current or past skills. In each category, there are some examples to help give you ideas. Fill in each section as accurately as possible. Include all activities even those that seem trivial. Even seemingly insignificant activities can display very positive and desirable personality traits. You will find that every time you review it, you will add new things. This will be a valuable reference which will help you convey your message in all of your applications, essays, and interviews. Consult the *Qualifications Form* sheet prior to completing your application, writing a resume, essay, or interview.

Before corresponding with sponsors for scholarship information, you must understand the importance of each letter, application, and essay. **They are sales tools.** Your application must sell your value(s) and demonstrate that you not only meet the qualification, but have the ability and desire to succeed in your chosen career.

Recall the activities you were involved in your freshman, sophomore, junior, and seniors years in high school or college years.

Your list may include:

School Activities
Office's you've held
Clubs
Team Sports
Honor Society
Newspaper / Yearbook
Chorus / Orchestra / Band

Outside Activities
Church Groups
Jobs
Hobbies
Tutoring
Sports
Other Languages

Way to Increase Your Qualifications
Extracurricular Activities are great ways to become a well-round individual and improve your chances of winning a scholarship. Most scholarship committees look for students who are involved in outside activities and have good grades. Devote free time to summer camps, newspaper editing, school tutoring, reading, athletics, summer and part-time jobs, community service, travel (trips with family and friends), etc. You can create productive activities which involves your friends. Your life will be enlighten by your participation. You will have an easier time writing essays and displaying your personality and character to the interviewers.

Your Goal is to....MAKE YOURSELF STAND ABOVE THE OTHER COMPETITION

Provided below is brief explanation of some activities to consider:

- **Take a CPR course** at your local Red Cross or fire department. The courses are usually free of charge and last about eight hours. This is an excellent course that is recommended for everyone. The benefits are tremendous for such a small investment of time. Your certification in CPR suggests: 1) your sense of social responsibility and desire for personal growth, 2) the ability to make quick decisions, and 3) awareness of life and death situation. (Skills such as this adds credentials / skills to your resume)
- **Do some volunteer and/or community work.** This shows your desire to give of yourself.
- **Take a position in an organization** or committee such as a fraternal organization, club, or group. As a member of a board or a committee, it shows your ability to work within an organizational structure.

These are just a few ways to quickly increase your qualifications. Use your imagination! Reread and "TAKE A LOOK AT YOURSELF. There are numerous activities that you can engage in which demonstrate that you possess the most desired traits.

After the completion of your *Qualifications Form*, you have created a reference sheet to use in all stages of the application process. You should review your qualifications sheet before completing an application.

Remember

Your Application and Essay are SALES TOOLS!!!

Writing An Effective Essay
The essay section is one of the most crucial parts of any scholarship application. Million of scholarships go unclaimed due to the reluctance of prospective applicants to write the essay(s) required for the applications. Writing an effective essay may be the most difficult process to receive scholarship funds. However, this is the most important part of the application. The reason this is very important is that without essays every application looks the same, with the same basic information (grades, financial need, extracurricular activities, etc.). The essay allows the candidate to express his/her ideas, convey personality, and exhibit the ability to write effectively. The essay is a key factor to whether a candidate is invited to an interview or awarded the scholarship.

Essays provides the scholarship committee information about your strengths, weakness, goals, dreams, attitude, personality, the way you think and how you possibly solve problems.

All essays have a basic structure. An essay begins with the introduction, body, and conclusion. The introduction has two primary functions. First, it is the foundation of your letter and it grabs the reader's attention. It serves as the outline in which you are about to cover in your essay. The body of the essay contains all of your ideas, facts, opinions, and supporting information. There should be a flow associated with each paragraph. Your ideas should transition from one another. The final paragraph is the conclusion. Your conclusion should summarize your letter in a few sentences and reinstate your main idea(s). It summarizes your statements in the body of the essay.

Tips For Writing An Effective Essay(s):

There are various ways to writing an essay. Provided below may help you organize your thoughts and provide you with some structure.

- Start writing your essay early. Don't want until the last minute.
- Identify what the essay question is asking you.
- Write down your ideas on a sheet of paper. Brainstorm!
- Organize your thoughts
- Make sure your essay covers the topic(s) asked in the question. Address every part of the required topic.
- Write a meaningful essay. Write about your genuine interest, something that is of great importance to you, that effects you, and hopefully will impact the person reading your essay. Include any of your achievements, qualifications, or skills that may be related to the sponsor's reason for awarding the scholarship.
- Talk to your academic advisor and others about the essay.
- Have several qualified people proofread your essay for grammatical errors and misspelled words. **AVOID** any errors!
- Carefully select your words. An essay should sound coherent and concise, not conversational.
- Allow a few days to write your essay
- Read your essay out loud. This will allow you to hear some of your mistakes in which your eyes may overlook.
- Read sentence by sentence to check to errors
- Make sure words are used properly
- Revise, reread, revise, reread, repeat
- Finally, mail the essay and application to scholarship sponsor a week before the deadline approaches. This gives you some time for errors or revisions. This also decreases the chances of your application and essay getting lost or delayed.

- Make sure that you are under the specified word count or page limit. There are several ways to correct the problem of going over the number of pages or words.

If this is the case, you should:
1. Adjust the margins
2. Reduce the typeface size (if the guidelines specifically request what margin size and typeface size..use them)
3. Reduce or rephrase some sentences and words

Actual Essay Questions From Scholarship Applications

The following questions have been selected from actual scholarship applications. You may be asked these questions on an application or interview. Answering the sample essay questions may help you prepare for the actual scholarship application(s). Use the Notes, located in the appendix section, to start writing your essay(s).

- What are your educational goals?
- How will this scholarship help you achieve your goals?
- What are your interests and/or experience in the aviation industry?
- Explain how your educational efforts will enhance your potential contribution to aviation.

- ✈ Why are you pursuing a career in aviation? (Explain your choice of aviation as a career path.)

Selecting Someone For A Recommendation Letter
Recommendation letters are another important part of your application, if it is requested. They allow the scholarship committee to form an idea of who and how you are as an individual thorough someone else's opinion. There are several things to consider when selecting someone for a recommendation. Consider the following tips:

- ✈ Choose someone that you have spent a considerable amount of time with. This individual should have an idea of you goals, accomplishments, potential, etc. People you should consider are your teachers, counselors, flight instructors, mentor, or current / past employer.
- ✈ Be sure to give your reference a considerable amount of time. Teachers, counselors, and others often get busier as the year continues. You want an excellent letter of recommendation. So give them time to write an excellent letter.
- ✈ Provide the reference person with some background materials. Knowing only about your academic performance will not win over the sponsors. An example would be your Extracurricular activities and awards outside of school, involvement in the community, goals, etc.
- ✈ Only ask for recommendations from people who know you and not those who are aquatinted with you.
- ✈ Your recommendation letter should include your academic performance, leaderships, goals, and outside activities.
- ✈ You may assist the reference along with your recommendation. Provide them with a list of things you want them to write about.
- ✈ Ask your reference for any ideas that would make you stand out above the other competition.
- ✈ Do not forget to send your reference(s) a Thank You letter or card.

Important Tips:

- ✈ Make photocopies of all materials you send out, then create a file for each scholarship.
- ✈ Make copies of completed applications, use them for future references.

What To Do After Mailing The Application
There are several ways to ensure that the sponsor has received your application.

- ✈ Mail the application via certified mail. (This will ensure that the organization will receive your application.)
- ✈ Send a self-addressed stamp along with your application and request the sponsor to mail it back to you. (This will be an indication that they have received your application.)
- ✈ Mail the application via first-class mail. After a few weeks, call the sponsor to make sure they received your application.

The only thing you can do after mailing your application is to be patient and wait for a decision. If you are not awarded a scholarship by a specific organization, **reapply next year**.

Persistence & Preparation are the Keys to Success.

What To Do If You Are Invited For An Interview By The Sponsor
Some local and regional scholarship sponsors may require the applicant to have an interview. This allows the interview committee to met the candidate face to face. This meeting will shuffle through the essays and exaggerated recommendation letters. This meeting is your time to shine and **SALE YOURSELF**.

If the interview is optional. It is recommended that you attend the interview. An interview always makes the difference, it will determine the outcome of the award. Preparing for an interview may be a difficult task to some students.

Provided below is a list of commonly asked questions during an interview. More specific question may be asked depending on the type of scholarship. If the scholarship is designated for a particular ethnic group, club affiliation, etc., be expected to answer questions relating to that area.

After reviewing the list, reread the questions one by one, and attempt to answer them. Write out the question(s) and answer(s) on a separate sheet of paper. Do not attempt to falsify any examples or create lies. Relate your actual experiences by incorporating them into your answer(s).

- What would you like to achieve in college?
- What do you see yourself doing after college?
- Why should we award you this scholarship?
- What are your favorite academic subjects? Why?
- Give some examples of how you have demonstrated leadership skills?
- What kind of activities are you involved in outside of school?
- What field are you planning to major? Why?
- What are your strengths and weaknesses?
- Why should we select you over the other applicants?
- Where do you see yourself five and/or ten years from now?
- What are your short and long-term goals?

If asked questions about your strengths, weakness, or qualifications; give a solid example of how each items helped you achieve your goal. You do not want to read off a list of accomplishments, qualifications, etc. Some of the interview questions may be similar to the sponsor's essay question(s).

Preparing For The Interview

Before you attend the interview, make sure you practice and have an idea of what to expect by reviewing the general questions. Have someone pretend to be an interviewer to critic your responses. This is considered to be a mock interview. Ask your teacher, counselor, parent, or friend to act as the interviewer. The mock interview should simulate the actual interview.

This feedback is very important to identfiying your possible weaknesses. Ask the "interviewer" a few questions about your answers, what you can do better, how do you look and sound, and any other questions you may have. You do not want to appear to be unorganized or unprepared during the interview. You should be able to provide a confident answer to each question asked. The Key is to PRACTICE!

For each question asked, you want to emphasize your good points. Think of how you can incorporate your positive character, goals, or qualities into your answer. Do not rush into answering the question. Provide a clear answer.

Here are a list of Do's and Don'ts during an interview.

Do's

1. Be yourself. Do not try to be someone you are not. Answer each question honestly and truthfully.
2. Incorporate your experiences into your answers..
3. Use a firm handshake. When you first meet the interviewer, shake his/her hand firmly. Avoid using a crushing or weak handshake.
4. Make eye contact. Throughout the interview, you should focus your eyes on the interviewer, but do not stare. Maintaining good eye contact shows that you are attentive, confident, and respectful. While shaking the interviewer's hand, make eye contact.
5. Sit facing the interviewer. You want to project that you are alert and attentive. Positioning your body away for the interviewer may give the interviewer the impression that you are rude. Maintain good posture, but try to be relaxed. Sit so that you are facing the interviewer, and avoid crossing your arms. This will make you seem defensive. You want to have an "open" appearance and interested at all times.
6. Smile and laugh when appropriate.
7. Show interest in the interviewer's background and organization. Do some research about the organization, scholarships guidelines, etc.
8. Bring a portfolio of your work, lists of activities, newspaper clippings, etc. Only bring items that will contribute to your application and interview.

Do not assume that the interviewer has reviewed your entire application. The interviewer may not know what activities you have been involved in, job experiences, etc. Bring a copy of your *Qualification Form* you completed. Provide a copy for the interviewer if it is requested and make sure it is typed.

9. Research the scholarship sponsor and prepare a list of questions for the interviewer.

Sample questions you may want to ask:

✈ What are the responsibilities of a recipient of this scholarship?

✈ What is the selection process?

✈ When should I expect a response?

✈ Is the scholarship renewable?

Don'ts

1. Do not dress casually. Do not show up in jeans, T-shirt, ear-rings, fancy jewelry, pagers, etc. Men should wear a conservative suit and tie. Women should wear a conservative dress or a pants suit, and do not use excessive makeup.
2. Do not arrive up late. It is recommended that you show up ten minutes early. This will give you time to relax. Showing up late does not give a good impression of you.
3. Do not talk too much about one particular thing. Answer the question and move on. Do not go into details unless asked by the interviewer.
4. Do not make inappropriate or rude comments. Avoid all of the following: profanity, jokes, personal comments about the interviewer, political or religious comments. Things of this nature may eliminate you from the list of eligible applicants.
5. Nervous habits to avoid: biting your nails, twiddling your thumbs, playing with a pen or pencil, shaking your leg, tapping your foot during an interview, etc. These habits and others may be distracting.
6. Avoid drifting off while the interviewer is speaking. Pay attention and listen to every word the interviewer says. Do not look out the window, at the furniture, or at the walls. Stay focused!
7. Words to avoid: "um," "like," or "you know", etc. Do not interrupt the interviewer when he/she is

speaking. If you need to consciously avoid those words, take a few moment before you respond and think about what you are about to say. Always try to provide a clear answer.

8. Do not forget to send a Thank You card or letter. Make sure you thank the interviewer for his/her time and consideration.

What To Do After Winning A Scholarship or Grant

If you won a scholarship from an outside source (organization, foundation, individual, etc.), it is very important to send a Thank You letter. Thank You letters are an effective way of showing your gratitude, and the organization's continuation to support other applicants pursuing a higher education or career goals.

A Thank You letter should be short, simple , and sincere. You can add your personal touch by hand-writing the letter.

An effective letter should contain the following elements:

- ✈ Date (This allows you to keep track of when the letter was sent.)
- ✈ Your name, address, phone number, e-mail address (as appropriate)
- ✈ Scholarship organization's name, coordinator (if known), address
- ✈ Gratitude for winning the scholarship
- ✈ Expressions of appreciation for their time and consideration
- ✈ "Thank You, " or "Sincerely," to close the letter
- ✈ Sign the letter before mailing it

Sample Thank-You Letter

(Date)

(Your name)
(Address)
(Phone Number)

(Scholarship Coordinator - if listed)
(Scholarship name)
(Scholarship Address)

Dear Scholarship Coordinator (use name if known),

I would like to thank (organization or sponsor's name) for awarding me (name of the scholarship). This award will assist me in achieving my career goals. Thank you for you time and consideration.

Sincerely,

(Sign Your Name)
(Print Your Name)

* *This is only an example of a Thank-You Letter.*

Notifying Your School After Winning An Outside Scholarship or Grant

After winning any outside scholarships or grants, you must inform your school's financial aid office. Your financial aid package may be adjusted to make room for outside sources of financial aid. This is because you are receiving financial assistance from the government and there are federal guideline to adhere too. These guidelines calculate how much financial assistance the institution can award the student. Most financial aid offices will reduced the least desirable forms of financial aid (which may be loans with a high interest rate, parent loans, or work study).

If you do not notify your school. They will eventually find out about your outside award. If this is case, they will make adjustments as necessary. Avoid unpleasant surprises of reduced aid by your school. Your priority is to focus on your studies and not worry about money. If you are unsure about how the federal financial aid and outside scholarships effect your financial aid package, contact your school's financial aid office to speak with a counselor.

Some scholarships will be sent to your school after classes have begun. Depending on the type of outside award you receive, you might have to make arrangements with the scholarship sponsor, or bursar's office of your school. Depending on your situation and documentation, your school may extend credit to you based on certain expectations of the award.

Remember, the scholarship or grant that you are awarded may be based on certain aspects of your enrollment. Some of these aspects may be the number of credit taken each semester/quarter, maintaining a certain grade point average, or enrollment in a certain career field/major.

Renewing Scholarships

Renewal procedures vary depending on the type of award you receive and the sponsor's guidelines. Some sponsors offer one-time awards only, while others offer continued support. Some sponsors that offer continued support automatically renew previous recipients as long as they abide by their guidelines (i.e. number of credit taken each semester/quarter, maintaining a certain grade point average, or enrollment in a certain career field/major). Use the knowledge you have gained from this book to your advantage and continue your scholarship search throughout your college years and/or flight training.

Do Not Limit Your Options And Always Plan Ahead.

THIS PAGE INTENTIONALLY LEFT BLANK

Scholarship Opportunities

Now that you are prepared with helpful tips from the previous sections on how to write a request letter, completing all necessary forms, and composing a winning essay. It is time for you to uncover all the scholarships, grants, fellowships, internships, and loans available to you. The awards listed are sponsored by private and public corporations, organizations, foundations, individuals, colleges/universities, and state & federal government agencies.

This section has been designed so you can quickly identify available funding by specific categories. Every piece of information is here that you'll need to decide if a scholarship is right for you: complete contact information, description, eligibility requirements, deadline dates, and the number awarded and amounts. In order to search efficiently, use the index system. This system is designed to help you match yourself with as many funding sources as possible. The index is organized into several categories:

- All Majors
- Aeronautics & Astronautics
- Aerospace Science & Technology
- Administration / Management
- Aviation Maintenance
- Avionics / Aircraft Electronics
- Fellowships
- Flight Training
- Military Affiliation
- Special Interest and Affiliation
- Women
- Loans

Within each section, listings are arranged alphabetically by the name of the scholarship. This arrangement is intended to facilitate your search through the listings. When referencing sources based on any given category, simply find a heading that applies to you.

Please Note

Students are strongly encouraged to read the descriptions carefully and pay particular attention to the various eligibility requirements <u>before</u> applying for awards. Apply only if you meet the eligibility requirements. Do not apply if you do not meet all of the qualifications for eligibility. If you are eligible, you should enclose a stamped, self-addressed #10 business size envelope when requesting details and applications. Deadline dates often fluctuate, so you must write for a current application for each year's deadline.

Some scholarship sponsors will not allow us to list their awards, because some applicants do not follow or read the requirements carefully. Sponsors receive many requests from applicants who do not qualify. This creates unnecessary work and expense for sponsors who must answer requests.

When applying for financial assistance, do not send photocopies or computer-generated letters to every source you find. Read each scholarship information carefully. Write or type an individual letter, **ONLY** if you meet the eligibility requirements. The letter should clearly define your qualifications.

<u>Important Tip:</u>

- Be sure to cross-reference yourself to all awards for which you qualify. If you are a double major, check in both categories.

ALL MAJORS

❒ THE AAAE SCHOLARSHIP PROGRAM
COTE
P.O. Box 2810
Cherry Hill, NJ 08034

Deadline: Contact Sponsor

Purpose: This scholarship is offered to a number of students with a junior class standing or higher, and will be enrolled in an aviation program and have a grade point average of 3.0 or higher.

Eligibility: Unrelated to membership in AAAE. The selection criteria used includes academic records, financial need, participation in school, community activities, work experience, and a personnel statement. Write for more information.

Award(s) & Amount(s): $1,000

❒ AIR TRAFFIC CONTROL SCHOLARSHIP PROGRAM
Air Traffic Control Association Scholarship Fund, Inc.
Contact: Suzette Mathews
2300 Clarendon Boulevard, Suite 711
Arlington, VA 22201
Tel. (703) 522-5717
Fax (703) 527-7251

Deadline: May 1

Description: Air Traffic Control Association (ATCA) scholarships are awarded to promising men and women who are enrolled in programs leading to a bachelors degree or higher in aviation related courses of study and to full-time employees engaged in advanced study to improve their skills in an air traffic control; or aviation discipline. Each year, ATCA will consider qualified half to full-time students as candidates for award of scholarships, each to be up to $2500. Each half to full-time student scholarship will be at least $1500. ATCA also will consider qualified full-time employees engaged in advanced (although not necessarily degree) study for award of scholarships up to $600. The numbers and amounts of scholarship awards will vary depending upon the qualifications, financial need and number of outstanding candidates who apply, but in all events the total amount of scholarships awarded will be limited to income generated by the fund, and one-third of the scholarship funds available, in any year will be reserved for full time employees. Money from the fund which is not awarded as scholarships in their category in a given year will be accumulated and reserved for scholarship awards in that category in future years.

The Air Traffic Control Association also maintains a separate fund from which educational scholarships are awarded to children of air traffic control specialists. The number and amount of scholarships will vary depending upon the qualifications, financial need and number of candidates who apply, but in all events the total amount of scholarships awarded will be limited to the income generated by the fund. Income from the fund not awarded as scholarships in any given year may be accumulated as additional capital, or paid out in scholarships in any subsequent year.

ATCA scholarships are reserved for U.S. citizens only, but will be awarded without regard to sex, race, religion or national origin. Scholarships must be used within four years of the date awarded. If the recipient should change academic major, resubmission of the intended program of study to the ATCA scholarship committee is required for continued reimbursement.

Eligibility:

❒ Half to Full-time Student Candidate Requirements:
U S Citizen; Enrolled (or accepted) in an accredited college or university and planning to continue the following year. Course work is related to his/her planned aviation-related career and leading to a bachelor's degree (or higher). Attendance 5 equal to at least half time (6 hours). Must have a minimum of 30 semester or 45 quarter hours still to be completed before graduation. (For Half to Full-time Students Only) A paper on the subject: "How My Education Efforts Will Enhance My Potential Contribution To Aviation." This paper should be typed, doubled spaced, 400 words maximum, and shall also address your financial need.

❒ Full-time Employee Candidate Requirements:
U.S. Citizen; Engaged in full-time employment in an aviation-related field in the Federal Government, U.S. military service or industry. Course work designed either for employees skill in an ATC or aviation discipline

❒ Children of Air Traffic Control Specialists
U.S. Citizen; Must be the child, natural or by adoption. of a person serving, or having served as an air traffic control specialist be it with the U.S. Government, U.S. Military, or in a private facility in the United States. Must be enrolled (or accepted) in an accredited college or university and planning to continue the following year. Attendance equal to at least half-time (6 hours). Must have a minimum of 30 semester or 45 quarter hours still to be completed before graduation. Course of study leads to a bachelors degree or higher.

Award(s) & Amount(s): Varies

❒ ALLEN H. AND NYDIA MEYERS FOUNDATION
P.O. Box 100
Tecumseh, MI 49286
Tel. (517) 423-7629

Deadline: March 15

Purpose: To provide grants to high school seniors planning on majoring in the physical sciences, engineering, aviation, or related fields. Must be residents of Lenawee County, MI .

Award(s) & Amount(s): 25, $350 - $500

❒ AVIATION COUNCIL OF PENNSYLVANIA
Aviation Council of Pennsylvania
3111 Arcadia Ave.
Allentown, PA 18103
Tel. (610) 797-1133
Fax: (610) 797-8238

Eligibility: **The applicant must be a Pennsylvania resident and funds must be used in Pennsylvania.**

Award(s) & Amount(s): Varies, (3 average, $1,000)

❒ AVIATION DISTRIBUTORS AND MANUFACTURERS ASSOCIATION
Contact: Charlotte Keyes
1900 Arch Street
Philadelphia, PA 19103-1498
Tel. (215) 564-3484
Fax (215) 564-2175 or (215) 963-9784

Deadline March 15

Description: ADMA will award two $1,000 scholarships to selected students each year. These will be selected from among the following: One $1,000 scholarship to a Bachelor of Science candidate with an Aviation Management major with emphasis in any of the following four areas: General Aviation, Airway Science Management, Aviation

Maintenance and Airway Science Maintenance Management; Bachelor of Science candidate with a major in Professional Pilot with any of the three following emphasis: General Aviation, Flight Engineer, or Airway Science A/C Systems Management. One $1,000 Scholarship to a second year student in an A&P (Aircraft and Powerplant) educational program (two year accredited aviation technical school.)

Eligibility: Applicant must be pursuing an aviation management or professional pilot degree. Applicant must complete a 1,000 word or more essay, describing themselves, their activities, goals and need for financial assistance. Applicant must submit two or more recommendations in writing by individuals from the applicant's school, aviation community, clergy or others. Applicant must submit transcripts with the application. Professional pilot major must also provide a progress report of their flight training. Professional pilot major must provide a copy of their: FAA medical certificate and pilot license.

Specific Requirements: Applicant must contact one of the ADMA members to request a Scholarship Application Form. Applicant must complete an ADMA Scholarship application.

Award(s) & Amount(s): Varies

❒ AVIATION INSURANCE ASSOCIATION SCHOLARSHIP

Aviation Insurance Association (AIA)
Contact: John Donica, AIA Exec. Dir., Scholarship Committee
PO Box 2966
Redmond, WA 98073
Tel. (425) 869-9522
Fax: (425) 861-6499

Deadline: September

Description: Open to all students enrolled in an aviation program at a college or university

Eligibility: 1) Must have completed at least 30 college credits, 15 must be in aviation. 2) must have a GPA 2.5 on a 4.0 scale or higher, 3) must be a U.S. citizen, 4) must submit a letter describing activities, indicating leadership qualities, goals, and reason for applying, 5) at least one letter of recommendation from an employer or instructor, 6) transcript must all accompany the application.

Award(s) & Amount(s): 1, $1,000

❒ BACCALAUREATE DEGREE COMPLETION PROGRAM (BDCP)

U.S. Navy
Attn: Naval Education and Training Program Management Support Activity
6490 Saufley Field Road
Pensacola, FL 32509-5204
Tel. (904) 452-1806
Toll Free (800) USA-NAVY

Deadline: Contact Sponsor

Purpose: To provide incentives to students from minority or educationally underprivileged backgrounds to continue their education and receive commissions in the Navy following graduation.

Eligibility: Students in junior college, community college, or 4-year college in any major who wish to serve in the Navy as officers following receipt of the Bachelor's degree. They must be under the age of 25 at the expected date of commissioning. Preference is given to minority students or those from an educationally underprivileged background.

Special features: Following graduation, participants in this program attend officer candidate school or aviation officer candidate school for 4 months and are commissioned as Navy ensigns. They have an active duty service obligation that varies from 4 to 7 years, depending upon specialization within the Navy. Further information is available from local Navy recruiters or Navy Recruiting Command, 801 North Randolph Street, Arlington, VA 22203-1991.

Note; Participants become active reserve enlisted Navy personnel and receive the pay of an E-3 seaman, or about $1,000 per month; the exact amount depends on the local cost of living and other factors. Duration: Until completion of a Bachelor's degree.

Award(s) & Amount(s): Varies each year.

❒ THE BILL FALCK MEMORIAL SCHOLARSHIP
Sponsored by EAA Chapter 474
Experimental Aircraft Association Foundation (EAA)
Scholarship Program
P. O. Box 3065
Oshkosh, WI 54903-3065
(414) 426-4888

Deadline: April 1 (depend on funding availability)

Eligibility: One scholarship awarded to an individual who has demonstrated a continuing quality in personal academic and aviation pursuits and may be applied toward the achievement of any aviation related formal education or training. Scholarship is not based on endowment funds and sometimes not awarded.

Award(s) & Amount(s): 1, $200.

❒ CHARLES H. GRANT SCHOLARSHIP
Academy of Model Aeronautics
Contact: Education Coordinator
5151 East Memorial Drive
Muncie, IN 47302
Tel. (765) 287-1256 ext. 272
Fax: (765) 289-4248

Deadline: Contact Sponsor

Description: The amount and number of scholarships to be presented for the current year will depend to a large extent the number of applicants and their qualifications. The scholarships will be distributed in various amounts on the basis of AMA modeling activities, scholastic achievement, and citizenship achievement.

Eligibility: **Applicant must be a member for 3 previous years (3rd year may be year of application).** The most desirable applicant is the one who is academically superior and also is a participant in many aspects of school, modeling, and the community. The applicants are rated in several major categories, including grade point average, and test results. High achievement in all of the categories is important for the maximum amount of scholarship awarded to an individual. A person who is not at the top of his/her class may win a significant scholarship award if he/she is active in the community, or in modeling, or both.

Award(s) & Amount(s): Several, up to $20,000

❒ CHARLIE WELLS MEMORIAL AVIATION SCHOLARSHIP
Charlie Wells Memorial Aviation Scholarship, Inc.
Contact: Jim Walker
1835 S. 4th St.
Springfield, IL 62703-3146

Deadline: August

Eligibility: Applicant must be a U.S. citizen attending a school in the U.S. Send SASE for application.

Award(s) & Amount(s): 2, $500

❒ **DONALD BURNSIDE MEMORIAL SCHOLARSHIP**
AOPA Air Safety Foundation
421 Aviation Way
P. O. Box 865
Frederick, MD 21701
Tel. (301) 695-2170

Deadline: March 31

Eligibility: Must be enrolled in and plan to continue a college curriculum leading to a degree in the field of aviation, be of sophomore standing at the time of application, have an overall grade point average of at least 2.5 on a 4.0 scale, and submit a 250 word (typed, double-spaced) paper on "Why I Wish to Pursue a Career in Aviation". Submit an official transcript with an application.

Award(s) & Amount(s): 1, $1,000.

❒ **EAA AVIATION ACHIEVEMENT SCHOLARSHIPS**
Experimental Aircraft Association Foundation (EAA)
Scholarship Program
P. O. Box 3065
Oshkosh, WI 54903-3065
Tel. (414) 426-4888

Deadlines: April 1 (Depend on funding availability)

Eligibility: Presented by the EAA to individuals active in sport aviation endeavors to further their aviation education or training. Scholarships are not based on endowment and are sometimes not awarded.

Award(s) & Amount(s): 2, $300

❒ **EDUCATIONAL COMMUNICATIONS SCHOLARSHIPS**
Contact: Educational Communications Scholarships
721 N. McKinley Rd
Box 5002
Lake Forest, IL 60045

Eligibility: For undergraduate study, criteria are test scores, GPA, essay, work experience. Must have taken SAT or ACT.

Award(s) & Amount(s): Varies

❒ **EUGENE S. KROPF SCHOLARSHIP**
Contact: Bernard W. Wulle
Aviation Technology Department
1 Purdue Airport
West Lafayette, IN 47906 - 3398

Deadline: May

Description: The University Aviation Association annually awards scholarship to applicants who are enrolled in an aviation-related curriculum at a UAA member college or university. Awards will be made to U.S. citizens without regard to gender, race, religion, or national origin.

Eligibility: Applicant be a U.S. citizen. Applicant must be enrolled in and plan to continue a college curriculum leading to a two-year or four-year degree in the field of aviation. Applicant must be officially enrolled in a UAA member institution in an aviation - related program. Cumulative undergraduate grade point average of 3.0 or above on a 4.0 scale. Completed application must be accompanied by a 250 - word, typed, double - spaced paper on “How Can I Improve Aviation Education.”

Award(s) & Amount(s): $500

❒ HAROLD S. WOOD SCHOLARSHIP
National Intercollegiate Flying Association (NIFA) and General Aviation Manufacturers Association (GAMA)
Ms. Bridgette Mikala
General Aviation Manufacturers Association
1400 K Street, N.W., Suite 801
Washington, D.C. 20005
Tel. (202) 393-1500

Deadline: February

Eligibility: Applicants must have at least a 3.0 GPA and be enrolled in an aviation curriculum. Selection will be based on collegiate aviation participation as well as community and social activities.

Award(s) & Amount(s): 1, $500

❒ HERBERT L. COX MEMORIAL SCHOLARSHIP
Experimental Aircraft Association Foundation (EAA)
Scholarship Program
P. O. Box 3065
Oshkosh, WI 54903-3065
Tel. (414) 426-4888

Deadline: April 1

Eligibility: Applicants must be accepted at or be attending a four-year accredited college or university in pursuit of a degree leading to an aviation profession. Must remain in good standing but GPA will not be more important than other criteria in making the selection. Must show unmet
financial needs for educational expenses.

Award(s) & Amount(s): 1, $800 or more (amount dependent on endowment interest)

❒ THE LANDRUM AND BROWN SCHOLARSHIP
Landrum & Brown and American Association of Airport Executives (AAAE)
The Landrum & Brown Scholarship, c/o AAAE
4212 King Street
Alexandria, VA 22302
Tel. (703) 824-0500 Ext. 26

Deadline: March 31

Purpose: To fund the educational pursuits of a worthy junior or senior undergraduate or a graduate student in an aviation program.

Eligibility: Applicants must be full time students, must be enrolled in an aviation program at an accredited college or university, preference given to juniors, seniors, and graduate students. Must have GPA of 2.75 or higher. Must submit an application with necessary documents. Sponsor is an airport planning consulting firm.

Award(s) & Amount(s): 1, $5,000

❒ MAPA SAFETY FOUNDATION, INC.
PO Box 460607
San Antonio, TX 78246 - 0607
Tel. (2100 525-8008
Fax (210) 525-8085

Deadline: September

Eligibility: Currently enrolled in a course of study that would promote general aviation safety. Applicant must have completed 50% of studies at the time of the awarding scholarship. Applicant must either be a MAPA member or sponsored by a MAPA member. Must have a minimum grade point average of 3.0/4.0 scale. Must provide the Board with a letter explaining how his/her future career would help promote general aviation safety and two letters of recommendation from his/her instructor.

Award(s) & Amount(s): 1, $2500

❒ MARINE CORPS HISTORICAL CENTER RESEARCH GRANTS
Marine Corps Historical Center
Building 58
Washington Navy Yard
Washington, DC 20374-0580

Purpose: To encourage graduate-level and advanced research in Marine Corps history and related fields.

Eligibility: While the program concentrates on graduate students, grants are available to other qualified persons as well. Applicants for the grants should have the ability to conduct advanced research in those aspects of American military history and museum activities directly related to the U.S. Marine Corps. Applicants may suggest research projects, or the staff of the Marine Corps Historical Center can provide guidance in selecting an appropriate topic. The proposed research may encompass such diverse topics as wars, institutions, organization and administration, policy, biography, civil affairs and civic action, civil-military relations, weaponry and technology, manpower, training and education, strategy, tactics, and logistics, as well as the interaction of diplomatic, political, economic, social, and intellectual trends affecting American military affairs during peace and war. Proposed research may also deal with such museum curatorial fields as exhibit design, military art, ordnance, uniforms, equipment, aviation, and other related topics. The program gives preference to projects covering the pre-1975 period. In all cases, the proposed research must result in a product that directly furthers or illuminates some aspect of the history of the Marine Corps. Examples of such products are an article for a professional journal, a publishable monograph or essay, a bibliography, a work of art, a museum display, or a diorama. Evaluation is based on ability, the nature of the proposed research, and the value of the research to the Marine Corps' historical program. All awards are based on merit, without regard to race, creed, color, or sex.

Limitations: Recipients are expected to do part of their research in Washington, D.C.

Award(s) & Amount(s): The number of awards vary each year. Grants range from $400 to $2,000. Funds are paid in 2 installments, half on the initiation of the approved project and the second half on its successful conclusion. There are no restrictions on how the recipients may use these funds.

❒ MCALLISTER MEMORIAL SCHOLARSHIP
AOPA Air Safety Foundation
421 Aviation Way
P. O. Box 865
Frederick, MD 21701
Tel. (301) 695-2170

Deadline: March 31

Awarded jointly annually by the AOPA Air Safety Foundation and the University Aviation Association. Must be enrolled in and plan to continue a college curriculum leading to a degree in the field of aviation, be of sophomore standing at the time of application, have an overall grade point average of at least 2.5 on a 4.0 scale, and submit a 250 word (typed, double-spaced) paper on "Why I Wish to Pursue a Career in Aviation". Submit an official transcript with application.

Award(s) & Amount(s): 2, $1,000

❐ MEA AVIATION SCHOLARSHIPS

10423 Fyfe Court,
Fairfax, VA 22032
E-mail: mea@aol.com.

Deadline: December 31

Description: Mid-Atlantic Educational Aviation has a scholarship available for basic or advanced education within an aviation or aerospace field. Fields may include flight instruction, air traffic control, aircraft maintenance, airport management, and aeronautical or aerospace engineering. Instruction may be at any location, but it must be provided by a state or federally licensed school, and it must be held at an accredited college, university or FAA-approved flight school. Applicants must be 16 or older.

Award(s) & Amount(s): $250

❐ MONTGOMERY GI BILL (ACTIVE DUTY)

Department of Veterans Affairs
810 Vermont Avenue, N.W.
Washington, DC 20420
Tel. (202) 233-4000
Toll Free(800) 827-1000

Purpose: To provide financial assistance for post-secondary education to new enlistees in any of the armed forces.

Eligibility: Eligible for this assistance are persons who enlist in the Army, Navy, Air Force, Marines, or Coast Guard after July 1, 1985, as well as persons who first perform full-time National Guard duty after November 29, 1989. Participants must serve continuously on active duty of enlistments for 3 years or longer, or for 2 years of an initial active duty obligation of less than 3 years. An individual may also qualify by serving 2 years on active duty followed by 4 years of Selected Reserve service. Following completion of their service obligation, participants may enroll in colleges or universities for associate, bachelor, or graduate degrees; in business, technical or vocational schools; for apprenticeships or on-job training programs; in correspondence courses; or, until September 30, 1994, in flight training.

Special features: Further information is available from local armed forces recruiters. This is the basic VA education program, referred to as Chapter 30, for veterans and military personnel who enter or have entered active duty since July 1, 1985. The comparable program for those whose service began earlier is the Veterans Educational Assistance Program (VEAP) for service prior to June 30, 1985. Service personnel eligible for those benefits as of December 31, 1989, who served on active duty without a break from October 19, 1984 to June 30, 1988 (to June 30, 1987 if followed by 4 years' service in the Selected Reserve) also qualify for this program, without contributing the $100 per month.

Duration: 36 months; active duty service members must utilize the funds within 10 years of leaving the armed services; reservists may draw on their funds while still serving.

Award(s) & Amount(s): Varies each year. Enlistees contribute $100 a month for the first year of service; at the completion of their service obligation, they are repaid their contribution plus additional funds from VA in 36 monthly payments of $400 per month ($325 per month for enlistments of less than 3 years).

❐ MONTGOMERY GI BILL (SELECTED RESERVE)

Department of Veterans Affairs
810 Vermont Avenue, N.W.
Washington, DC 20420
Tel. (202) 233-4000
Toll Free (800) 827-1000

Deadline: Applications may be submitted at any time.

Purpose: To provide financial assistance for post-secondary education to reservists in the armed services.

Eligibility: Eligible to apply are members of the Reserve elements of the Army, Navy, Air Force, Marine Corps, and Coast Guard, as well as the Army National Guard and the Air National Guard. To be eligible, a reservist must 1) have a 6-year obligation to serve in the Selected Reserves signed after June 30, 1985 (or, if an officer, to agree to serve 6 years in addition to the original obligation); 2) complete Initial Active Duty for Training (IADT); 3) meet the requirements for a high school diploma or equivalent certificate before completing IADT; and 4) remain in good standing in a drilling Selected Reserve unit.

Financial data: Reservists who enlisted prior to June 30, 1985 can receive benefits for undergraduate degrees or for technical courses leading to certificates at colleges and universities. Reservists whose 6-year commitment began after September 30, 1990 may also use these benefits for business, technical, vocational, cooperative apprenticeship, on-job, correspondence, independent study, tutorial assistance, or (until September 30, 1994) flight training. Payments are made monthly. The rate for full-time study is $190 per month.

Special features: This program is frequently referred to as Chapter 106. Reservists who are enrolled for three-quarter or full-time study are eligible to participate in the work-study program.

Limitations: Benefits end 10 years from the date the reservist became eligible for the program. The VA may extend the 10-year period if the individual could not train because of a disability caused by Selected Reserve service. Certain individuals separated from the Selected Reserve due to downsizing of the military between October 1, 1991 and September 30, 1995 will also have the full 10 years to use their benefits.

Award(s) & Amount(s): Varies each year. Duration: Up to 36 months.

❑ NATIONAL AGRICULTURAL AVIATION ASSOCIATION
Contact: Deborah Russell
Tel. (202) 358-0935

Note: $1,000 educational scholarship will be awarded to a winner.

❑ NATIONAL BUSINESS AIRCRAFT ASSOCIATION SCHOLARSHIPS
National Business Aircraft Association
1200 18th St., N.W., Suite 400
Washington, D.C. 20036-2506
Tel. (202) 783-9000
Fax: (202) 331-8364
Website: http://www.nbaa.org
E-mail: info@nbaa.org

Deadline: October

Description: The National Business Aviation Association annually award five $1,000 scholarships to applicants who are studying aviation-related curriculums. Awards will be made to US citizens without regard to race, sex, religion, or national origin.

Eligibility: Applicant requirements are that the individual be a US citizen and must be enrolled in an aviation-related program at a NBAA and University Aviation Association (UAA) member institution. Must be a sophomore, junior or senior who will be continuing in school the following academic year and must have a cumulative GPA of 3.0 or higher. An official transcript from the applicants college or university must be submitted with the NBAA Scholarship application form. Application requires a 250-word paper describing the individual's interest in a career in aviation. A letter of recommendation from a member of the aviation department faculty at the institution in which you are currently enrolled. A resume must be submitted

Award(s) & Amount((s): 5 scholarships, $1,000

❐ PIONEERS OF FLIGHT SCHOLARSHIP
National Air Transport Association Foundation
Citizens' Scholarship Foundation of America, Inc.
P.O. Box 297
St. Peter, MN 56082
Tel. (507) 931-1682

Deadline: November 15

Eligibility: Applicants must be nominated by members of the National Air Transport Association (NATA) and be of sophomore or junior standing with a strong interest in pursuing a career in general aviation

Award(s) & Amount(s): 4, $2,500 (renewable)

❐ THE RICHARD LEE VERNON AVIATION SCHOLARSHIP
Sponsored by EAA and the Richard Lee Vernon Family
Experimental Aircraft Association Foundation (EAA)
Scholarship Program
P. O. Box 3065
Oshkosh, WI 54903-3065
Tel. (414) 426-4888

Deadline: April

Eligibility: Applicants must be accepted in a course of study in a recognized professional aviation training program in an institution of higher learning or aviation technical school. Must have demonstrated the ability to complete the course of training, attain acceptable grades and show financial need. May be applied toward the achievement of any aviation-related formal education or training.

Award(s) & Amount(s): Contact Sponsor

❐ SAFE ASSOCIATION SCHOLARSHIP
SAFE Association
Scholarship Committee
Embry-Riddle Aeronautical University
3200 North Willow Creek Road
Prescott, AZ 86301-8663.

Deadline: June 1

Eligibility: Must be a full-time student enrolled in a college degree pertaining to the field of safety and survival. Submit an application, a college transcript and recommendation forms and a narrative essay on a safety or survival topic. Not limited to aviation degrees.

Award(s) & Amount(s): 1, $1,000

❐ SILVER DART AVIATION HISTORY AWARD
Canadian Aviation Historical Society
National Headquarters
P.O. Box 224, Sta. A
Willowdale, ON, Canada M2N 5S8
Tel. (416) 488-2247

Deadline: March 15

Eligibility: The author of the best essay will be awarded the Silver Dart Aviation History Award, a scroll, and a $500 cash payment.

❒ TELEDYNE CONTINENTAL AVIATION EXCELLENCE SCHOLARSHIP
Sponsored by EAA and Teledyne Continental Motors
Experimental Aircraft Association Foundation (EAA)
Scholarship Program
P. O. Box 3065
Oshkosh, WI 54903-3065
Tel. (414) 426-4888

Deadline: April 1 (depends on funding availability)

Eligibility: Based on excellence in personal and aviation accomplishments and based on individual's potential to become a professional in any field of aviation. Not based on endowment funds and sometimes not available.

Award(s) & Amount(s): 1, $500.

❒ TRANSPORTATION CLUBS INTERNATIONAL SCHOLARSHIPS
Attn: Gay Fielding
P. 0. Box 52
Arabi, LA 70032

Deadline: April

Eligibility: Any student enrolled in an educational program in an accredited institution of higher learning, offering courses in Transportation, Logistics, Traffic Management, or Related Fields, who intends to prepare for a career in these areas. The awards will be based upon scholastic ability and potential, professional interest, and character. Financial need will also be given consideration.

❒ Hooper Memorial Scholarship(s)
Description: $1,500 TCI's first scholarship, awarded in memory of the fonder of the first Women's Transportation Club in the U.S. (Los Angeles, CA)

❒ Charlotte Woods Memorial Scholarship
Description: $1,000 Awarded in memory of a former TCI director who was responsible for the proclamation of the now annual "National Transportation Week". Student must be a TCI member, or a dependent of a member.

❒ Texas Transportation Scholarship
Description: $1,000 Awarded in memory of Gene Landis of Houston, Texas, to a student who has been enrolled in a school in Texas during some phase of their education (elementary, secondary, or high school).

❒ Ginger & Fred Deines Mexico Scholarships
Description: 2, $500 & $1,000 Awarded in memory of Ginger and Fred Deines, to a student of Mexican nationally and enrolled in a school in Mexico or US. Mr. Deines was a past International President of TCI.

❒ Denny Lydic Scholarship
Description: $500 Awarded in appreciation and recognition for his dedication to the field of transportation. He is a past International President of TCI and Continues to be very active and supportive.

Award(s) & Amount(s): Varies

❒ USAIG PDP SCHOLARSHIP
US Aircraft Insurance Group
NBAA PDP Scholarship
Contact: Annie Brown
1200 Eighteenth St., NW, Suite 400
Washington, DC 20036
Tel. (202) 783-9000
Fax: (202) 331-8364
Website: http://www.nbaa.org

Deadline: August

Description: The US Aircraft Insurance Group (USAIG) annually awards three $1000 scholarships to applicants who are enrolled full-time in a college or university offering the NBAA Professional Development Program (PDP). Awards will be made to US citizens without regard to sex, race, religion or national origin.

Eligibility: Applicant requirements are that the individual be a US citizen and must be enrolled in an aviation-related program at a NBAA and University Aviation Association (UAA) member institution. NBAA/UAA member college and universities offering PDP curricula are Embry-Riddle Aeronautical University, Central Missouri State University, Mercer County Community College, Purdue University, and University of North Dakota. Contact the department head or advisor for the institution's NBAA or UAA membership number.

Must be a sophomore, junior or senior who will be enrolled in an aviation-related two-year, four-year, or postgraduate degree program that incorporates the NBAA PDP. Must have proof of enrolled prior to distribution of the award. Must have a cumulative GPA of 3.0 or higher on a 4.0 scale. An official transcript from the applicants college or university must be submitted with the NBAA Scholarship application form. Application requires a 250-word paper describing the individual's interest and goals for a career in aviation. A letter of recommendation from a member of the aviation department faculty at the institution in which you are currently enrolled. A resume must be submitted. Scholarships will be made payable to the school.

Award(s) & Amount(s): 3, $1,000

❒ WILFRED M. POST JR. AVIATION SCHOLARSHIP
American Association of Airport Executives (AAAE)
Executive Secretary
American Association of Airport Executives
Northeast Chapter
4224 King Street
Alexandria, VA 22302
Tel. (703) 824-0500

Deadline: March

Eligibility: Applicants must be of junior or senior standing. Applicants must be studying for an aviation related undergraduate degree. Applicants must complete an application form.

Award(s) & Amount(s): 4, $1,000

AERONAUTICS / ASTRONAUTICS

❒ AIAA/INDUSTRY SCHOLARSHIPS
American Institute Of Aeronautics And Astronautics
Director of Student Programs 370 L'Enfant Promenade, S.W.
Washington, D.C. 20024-2518
Tel. (202) 646-7458

AIAA/Student Programs
1801 Alexander Bell Drive
Suite 500
Reston, VA 20191
Tel. (703) 264-7500
Fax (703) 264-7551
http://www.aiaa.org

Deadlines: February 1

Description: The objective of the American Institute of Aeronautics and Astronautics is to advance the arts, sciences, and technology of aeronautics and astronautics. The Institute encourages original research, furthers dissemination of new knowledge, fosters the professional development of those engaged in scientific and engineering activities, improves public understanding of the profession and its contributions, fosters education in engineering and science, promotes communication among engineers and scientists as well as other professional groups, and stimulates outstanding professional accomplishments.

❒ The Abe M. Zarem Award for Distinguished Achievement

Deadlines: Abstract Submittal Deadline - January 31; Deadline for Completed Papers - April 1; Announcement of Winners - June 24.

Description: The Zarem Graduate Student Awards have been established by AIAA and Dr. Abe Zarem as a means for students pursuing advance degrees (master's level candidates) in aeronautics and astronautics.

Eligibility: All master's level AIAA student members in good standing are encouraged to participate. To enter, the student must submit a writing of technical work (research paper) done at the master's level. Technical papers that are used as part of a research thesis are eligible and encouraged for the competition. Papers prepared for the Regional Student Paper Conferences may also be entered. Five copies of the paper must be submitted; each must bear the signature and the student number of the author. The paper must also bear the signature of the faculty advisor sponsoring the research. The author may enter the paper in the aeronautics category or the astronautics category. The papers are judged on technical merit. Criteria for judging of the papers are as follows: technical content, originality, practical application, style and form. Judging of the papers will be handled by the AIAA Student Activities Committee. Only single authored papers are permitted. The top paper in each category will be the Zarem Award Winner. The winners will receive medals and certificates at the AIAA Aerospace Sciences Meeting. The winner in the aeronautics category will be supported to the International Council of the Aeronautical Sciences Meeting, and the winner in the astronautics category will be supported to the International Astronautical Federation Meeting to present his/her paper.

Restrictions: Anyone involved in an AIAA National Technical Committee, Education Committee, or other AIAA National Member Committee, or National Subcommittee is not eligible for this scholarship.

Award(s) & Amount(s): Varies

❒ AIAA FOUNDATION UNDERGRADUATE SCHOLARSHIP PROGRAM

Student Programs
1801 Alexander Bell Drive
Suite 500
Reston, VA 20191-4344

Deadlines: Application requests: January 15; Submission of applications for the academic year: January 31

Description: This program was instituted in the fall of 1977 and now presents thirty scholarship awards annually as follows, provided worthy candidates are available: $2,000 each to one or more college sophomores, $2,000 each to one or more college juniors, $2,000 each to one or more college seniors.

Note: It is further planned to perpetuate this scholarship award by continuing the $2,000 yearly awards for the deserving sophomore and junior recipients (until completion of their senior year), and by presenting additional new awards each year as funds permit, to deserving college sophomores, juniors, and seniors.

Eligibility: The following eligibility requirements shall be met by applicants for the AIAA Foundation Undergraduate Scholarship Awards: Applicant must have completed at least one (1) academic quarter or semester of full-time college work. Applicant must have a college grade point average of not less than the equivalent of a 3.0 on a 4.0 scale. Applicant shall be enrolled in an accredited college or university. Applicant does not have to be an AIAA student member in good standing to apply, but must become one before receiving a scholarship award. Applicant's scholarship plan shall be such as to provide entry into some field of science or engineering encompassed by the technical activities of AIAA. Applicant shall not have, or subsequently receive, any other scholarship award which, when combined with the AIAA Foundation award, would provide a stipend greater than their tuition plus educational expenses (such as books, lab fees, etc.) estimated by the educational institution he/she plans to attend.

Sophomores and juniors who receive one of these awards are eligible for yearly continuation of the awards (until completion of their senior year) provided they maintain at least the equivalent of a 3.0 on a 4.0 scale, have excellent references, and submit well defined aerospace/aeronautical career goals. It should be stressed that regardless of GPA, renewal is not automatic. To apply for renewal, an individual must submit a completed application, career essay, official transcript and two (2) letters of recommendation. At least one of these letters must be from a professor from the student's university or college. Applicants must be either U.S. citizens or permanent residents of the United States.

Award(s) & Amount(s): Varies

❒ AIAA FOUNDATION GRADUATE AWARDS

1801 Alexander Bell Drive
Suite 500
Reston, VA 20191-4344

Deadlines: Application requests: January 15; Submission for applications January 31

Description: The American Institute of Aeronautics and Astronautics Foundation, in conjunction with AIAA Technical Committees (TCS), is seeking applications for the AIAA Foundation Graduate Awards. These awards are intended for graduate-level students who indicate by their approved department research an interest in any of the areas listed on page two of this application. The purpose of this program is to provide awards to qualified students actively participating in these specific research endeavors as part of their graduate studies. These endeavors include master's or doctoral thesis research and master's (non-thesis option) research projects.

Note: The award program provides $5,000 awards annually. College graduate students, pursuing Master of Science or Doctor of Philosophy degrees and conducting university department approved research associated with the specific disciplines outlined on the following page, are eligible. These awards are nonrenewable; that is, a selected candidate may receive the award only once.

Eligibility: The following eligibility requirements should be met by applicants for the AIAA Foundation/Graduate Awards: Applicant must have completed at least one (1) academic year of full-time graduate college work. Applicant must have in place, or underway, a university department approved thesis or research project specializing in one of the listed technical areas. Applicant cannot be a previous winner of this award. If the applicant has previously applied, but was not selected, then he/she may reapply. A postdoctoral candidate is not eligible. Applicant must have a graduate grade point average of not less than the equivalent of a 3.0 on a 4.0 scale. Applicant shall be enrolled in an accredited college or university graduate program under one of the related technical courses of study. Applicant's graduate study program plan shall be in support of the field of science and engineering encompassed by a specialized area. Applicant does not have to be an AIAA student member in good standing to apply, but must become one before receiving an award. Candidate's application package must be endorsed and/or cosigned by his/her Graduate advisor and the appropriate University Department Head. Applicant must be a citizen of the United States.

Areas of Specialization

❒ Gordon C. Oates Air Breathing Propulsion Graduate Award
Description: This program was instituted in 1985 by the Air Breathing Propulsion Technical Committee in honor of C. Oates. Eligible applicants will be participating in research endeavors in Air Breathing Propulsion as part of their graduate studies. Official recognition and presentation of the check to the recipient will be made in conjunction with the Awards Presentation Ceremony at the AIAA/SAE/ASME/ASEE Joint Propulsion Conference. In addition, the AIAA will present the recipient with a special recognition plaque in honor of this award. One or more Graduate Master of Science and/or Doctor of Philosophy Fellowships of $1,000 are awarded.

❒ The Liquid Propulsion Graduate Award
Description: This program was instituted in 1988 by the Liquid Propulsion Technical Committee. Eligible applicants will be participating in research endeavors in Liquid Rocket Propulsion and its related disciplines. Official recognition and presentation of the check to the recipient will be in conjunction with the Awards Presentation Ceremony at the AIAA/SAE/ASME/ASEE Joint Propulsion Conference. In addition, the AIAA will present the recipient with a special recognition plaque in honor of this award. Please note - this is a $1,000 Graduate Award funded by the Liquid Propulsion Technical Committee. One or more Graduate Master of Science and/or Doctor of Philosophy Fellowships of $1,000 are awarded.

❒ **The Martin Summerfield Propellants and Combustion Graduate Award**
Description: Funding for this award is provided through individual memorial gifts (given after the death of Martin Summerfield) and the AIAA Foundation. Eligible applicants will be actively participating in research endeavors in propellants and combustion as part of their graduate studies. Official recognition will be made in conjunction with the Awards Presentations Ceremony at the AIAA/SAE/ASME/ASEE Joint Propulsion Conference. In addition, the AIAA will present the recipient with a special recognition plaque in honor of this award.

❒ **Open Topic**
Description: In order to allow for the greatest possible flexibility in research, the AIAA Foundation has made available up to and including five (5) "non-designated" graduate awards. Eligible applicants must be actively participating in research endeavors in one of the more than 65 specialty areas represented by AIAA Technical Committees. (Be sure to indicate on the application your chosen area of specialization.) Official recognition and presentation of the check to the recipient will be made in conjunction with the Awards Presentations Ceremony at the AIAA Aerospace Sciences Meeting. In addition, the AIAA will present the recipient with a special recognition plaque in honor of this award.

❒ **The William T. Piper, Sr. General Aviation Systems Graduate Award**
Description: Funding for this award is provided by endowment gifts from the Piper Aircraft Corporation and the W.T. Piper, Sr. Foundation, in the name of William T. Piper, Sr. combined with support from the General Aviation Systems Technical Committee and the AIAA Foundation. Eligible applicants will be actively participating in research endeavors in General Aviation as part of their graduate studies. Official recognition will be made in conjunction with the Awards Presentations Ceremony at the AIAA Aerospace Sciences Meeting. In addition, the AIAA will present the recipient with a special recognition plaque in honor of this award.

Award(s) & Amount(s): Varies

❒ **ALLEN H. MEYERS SCHOLARSHIP FOUNDATION**
P.O. Box 100
Tecumseh, MI 49286
Tel. (517) 423-7629

Deadline: March 15

Eligibility: Awards are made once to any individual and are only renewable under unusual circumstances.

Award(s) & Amount(s): Variable

❒ **AMERICAN SOCIETY OF NAVAL ENGINEERS SCHOLARSHIP PROGRAM**
The American Society of Naval Engineers
Contact: Dennis Pignotti
1452 Duke Street
Alexandria, VA 22314-3458
Tel. (703) 836-6727
Fax (703) 836-7491
http://www.jhuapl.edu/ASNE
E-mail: asnehq.asne@mcimail.com

Deadline: February 15

Description: The American Society of Naval Engineers encourages college students to enter the field of naval engineering, and to provide support to naval engineers seeking advanced education in this field. Naval engineering includes aeronautical engineering, the design, construction and repair of ships and their installed systems and equipment, as well as research, logistics support, and the management of acquisition and maintenance.

Eligibility: The candidate will be applying for support for the last year of a full-time or co-op undergraduate program or one year of full-time graduate study leading to a designated engineering or physical science degree in an accredited college or university. A scholarship will not be awarded to a doctoral candidate or to a person already having an advanced degree. The candidate must be a United States citizen. The candidate must have demonstrated or expressed a genuine interest in a career in naval engineering, e.g. a student membership in a professional engineering society, extra-curricular engineering activities, etc. All graduate student candidates must be members of ASNE or SNAME.

Note: Selection criteria will be based on the candidate's academic record, work history, professional promise and interest in naval engineering, extra-curricular activities, and recommendations of college faculty, employers, and other character references. Financial need may also be considered. The award may be used for payment of tuition, fees, and expenses for students who meet the requirements listed.

Award(s) & Amount(s): 15 - 18; $2,500 per year for undergraduate students and $3,500 per year for graduate students

---

❒ ELECTRONIC INDUSTRIES FOUNDATION SCHOLARSHIP
1901 Pennsylvania Ave., N.W., Suite 700
Washington, D.C. 20006
Tel. (202) 955-5810

Deadline: February 1

Eligibility: Scholarships awarded to disabled high school seniors, undergraduate, or graduate students. Must be pursuing a career in aeronautics, computer science, engineering technology, etc.

Award(s) & Amount(s): 6 scholarships, $2,000

❒ SILVER DART AVIATION HISTORY AWARD
Canadian Aviation Historical Society
National Headquarters
P.O. Box 224, Sta. A
Willowdale, ON, Canada M2N 5S8
Tel. (416) 488-2247

Deadline: March 15

Note: The author of the best essay will be awarded the Silver Dart Aviation History Award, a scroll, and a $500 cash payment.

❒ U.S. AIR FORCE ROTC (3 AND 4 YEAR SCHOLARSHIP PROGRAM)
Contact: Capt. Teresa Kohlbeck
Tel. (612) 962-6330

Deadline: Contact Sponsor

Description: Over 4,000 scholarships awarded to high school seniors per year. To be used at any campus which offers Air Force ROTC. Scholarships are merit based, do not consider financial need, and pay tuition only. The fields include: Aeronautics, Aerospace, Astronautical Science, Civil Engineering, Mechanical Engineering, Mathematics Physics and more.

Eligibility: Applicant must be a U.S. citizen at least 17 years of age and under the age of 25 at the time of graduation from college. Furnish SAT or ACT scores, high school transcripts and record of extracurricular activities. Must qualify on Air Force medical examination.

Award(s) & Amount(s): Contact Sponsor

❒ UNITED STATES SPACE FOUNDATION
Education Department

2860 S. Circle Dr., Suite 2301
Colorado Springs, CO 80906-4184
Tel. (719) 576-8000
Fax (719) 576-8801

Note: All expenses except personal spending money. Differ each year

---

❐ VERTICAL FLIGHT FOUNDATION SCHOLARSHIPS
Vertical Flight Foundation
Attn: Scholarship Committee
217 North Washington Street
Alexandria, VA 22314
Tel. (703) 684-6777
Fax: (703) 739-9279
E-mail: ahs703@aol.com

Purpose: To provide financial assistance to college students interested in preparing for an engineering career in the helicopter or vertical flight industry.

Eligibility: Applicants must be full-time undergraduate or graduate students at an accredited school of engineering. They need not be a member or relative of a member of the American Helicopter Society. Selection is based on academic record, letters of recommendation, and career plans.

Special features: The Vertical Flight Foundation was founded in 1967 as the philanthropic arm of the American Helicopter Society.

Award(s) & Amount(s): 1 or more each year. Up to $2,000 per year. Duration: 1 year.

❐ WILLIAM T. PIPER, SR. GENERAL AVIATION SYSTEMS GRADUATE AWARD
American Institute of Aeronautics and Astronautics
Attn: Graduate Awards
370 L'Enfant Promenade, S.W.
Washington, DC 20024-2518
Tel. (202) 646-7400
Fax: (202) 646-7508

Deadline: January

Purpose: To provide financial assistance for graduate research on general aviation.

Eligibility: This program is open to graduate-level students who are actively participating in research on general aviation as part of their graduate studies. They may be working on a master's thesis, a doctoral dissertation, or a master's (non-thesis option) research project. Applicants must have completed at least 1 academic year of full-time graduate work, have earned at least a 3.0 grade point average, and be a U.S. citizen. Selection is based on the quality of the research proposal, the academic program being pursued, career goals, and recommendations. Applicants need not be a member of the American Institute of Aeronautics and Astronautics, but must become a member before receiving an award.

Award(s) & Amount(s): The stipend is $1,000. Funds are paid directly to the recipient. Duration: 1 year; nonrenewable.

AEROSPACE SCIENCE

❒ ALLEN H. MEYERS SCHOLARSHIP FOUNDATION

P.O. Box 100
Tecumseh, MI 49286
Tel. (517) 423-7629

Deadline: March 15

Purpose: Awards are made once to any individual and are only renewable under unusual circumstances.

Award(s) & Amount(s): Varies

❒ DR. ROBERT H. GODDARD SPACE SCIENCE & ENGINEERING SCHOLARSHIPS

National Space Club
655 15th St., N.W., #300
Washington, D.C. 20005
Tel. (202) 639-4210

Deadline: January

Purpose: These scholarships are awarded to college junior and seniors for study leading to increased knowledge of space research and exploration. Must plan to pursue a career in aerospace sciences and technology.

Award(s) & Amount(s): $7,500

❒ I.T UNDERGRADUATE RESEARCH ASSISTANTSHIP PROGRAM

Contact: Donna Rosenthal
Tel. (612) 628-8000

Deadline: No application deadline

Description: One award per scholarship is given per year. There is no application deadline. Selection taken from incoming freshman applicants to the Institute of Technology.

Eligibility: These Scholarships and fellowships are awarded to incoming freshman who have designated Aerospace Engineering and Mechanics as their projected major and who have indicated the I.T Honers application and interest in receiving an I.T Undergraduate Research Assistantship. Decisions are based on the complete University of Minnesota application and on the I.T Honors application.

Note: All awards provide the recipients the opportunity to participate in laboratory or other research in aerospace Engineering and Mechanics under the guidance of a faculty sponsor. For the Minnesota Space Grant Consortium Fellowship, the recipients must be a U.S Citizen.

Award(s) & Amount(s): $2000.

❒ MONTANA SPACE GRANT CONSORTIUM

Contact: Laurie Howell
Montana Space Grant Consortium
229A AJM Johnson Hall
MSU-Bozeman
Bozeman, MT 59717
TEL. (406) 994-4223

Deadline: April

Description: Grants are made on a competitive basis to students enrolled in fields related to the aerospace sciences and engineering. Examples of related fields include biological and life sciences, chemistry, geological sciences, physics and astronomy, mechanical engineering, chemical engineering, electrical engineering, computer sciences and civil engineering.

Eligibility: Applicants must be a U.S. citizen enrolled as a full-time student at a campus belonging to the Montana Consortium. Awards are for one year, but are renewable on a competitive basis.

Award(s) & Amount(s): $1,000 undergraduate scholarships and $10,000 graduate fellowships are received by the spring of each year.

❒ R. MINKIN AEROSPACE ENGINEERING SCHOLARSHIP
Contact: Donna Rosenthal
Tel. (612) 625-8000

Description: The Aerospace Engineering and Mechanics Award Committee reviews each eligible undergraduates GPA and makes their selection annually.

Eligibility: Open to a student majoring in Aerospace Engineering and Mechanics. It is awarded to a sophomore student with the best GPA after five consecutive quarters, with at least 72 credits completed satisfactorily.

Award(s) & Amount(s): $800.

❒ U.S. AIR FORCE ROTC (3 AND 4 YEAR SCHOLARSHIP PROGRAM)
Contact: Capt. Teresa Kohlbeck
Tel. (612) 962-6330

Deadline: Contact Sponsor

Description: Over 4,000 scholarships awarded to high school seniors per year. To be used at any campus which offers Air Force ROTC. Scholarships are merit based, do not consider financial need, and pay tuition only. The fields include: Aeronautics, Aerospace, Astronautical Science, Civil Engineering, Mechanical Engineering, Mathematics Physics and more.

Eligibility: Applicant must be a U.S. citizen at least 17 years of age and under the age of 25 at the time of graduation from college. Furnish SAT or ACT scores, high school transcripts and record of extracurricular activities. Must qualify on Air Force medical examination.

Award(s) & Amount(s): Contact Sponsor

❒ UNDERGRADUATE/GRADUATE ENGINEERING SCHOLARSHIPS
American Helicopter Society Vertical Flight Foundation
TEL. (703) 684-6777

Eligibility: Open to undergraduate students and graduate students in the following fields: Mechanical Engineering, Electrical Engineering, and Aerospace Engineering for study at accredited institutions in the U.S.A..

Award(s) & Amount(s): 8-9, $2,000

ADMINISTRATION / MANAGEMENT

❒ AAAE SCHOLARSHIP
American Association of Airport Executives (AAAE)
4212 King Street
Alexandria, VA 22302
Tel. (703) 824-0500

Deadline: May 15

Eligibility: Limit of 1 student scholarship application per university. Must apply through your university. Must have reached junior standing or higher, through graduate school. Cumulative GPA 2.75 on 4.0 scale or equivalent. Selection based on scholastic achievement, financial need and community activities. Must submit transcripts.

Award(s) & Amount(s): 10, $1,000

❒ ACI-NA COMMISSIONERS ROUNDTABLE SCHOLARSHIP
Airports Council International - North America
College of Technical Careers
Attn. Aviation Management and Flight Department
Southern Illinois University at Carbondale
Carbondale, IL 62901
Tel. (618) 453-8898

Deadline: November 1

Purpose: The primary purpose of the Airports Council International - North America Commissioners Roundtable Scholarship is to provide financial support to students working toward a career in Airport Management or Airport Administration.

Eligibility: To be considered you must: 1) Be officially enrolled in an accredited college or university in an undergraduate program focused on a career in Airport Management. 2) Possess a 3.0 GPA on a 4.0 scale or the equivalent. 3) Must reside and attend school in the U.S., Canada, Saipan, Bermuda, U.S. Virgin Islands or Guam.

Award(s) & Amount(s): Up to 4, $2,500

❒ ADMA SCHOLARSHIP PROGRAM
Aviation Distributors and Manufacturers Association
Charlotte Keyes
1900 Arch Street
Philadelphia, PA 19103-1498
Tel: (215) 564-3484
Fax: (215) 564-2175 or (215) 963-9784

Deadline: March 15

Description: The ADMA Scholarship Fund was established to provide assistance to students pursuing careers in the aviation field. ADMA will award two $1,000 scholarships to selected students each year. These will be selected among the following: One $1,000 scholarship to a Bachelor of Science candidate with an Aviation Management major with emphasis in any of the following four areas: General Aviation, Airway Science Management, Aviation Maintenance and Airway Science Maintenance Management; or, a Bachelor of Science candidate with a major in Professional Pilot with any of the three following emphasis: General Aviation, Flight Engineer, or Airway Science A/C Systems Management; or, one $1,000 Scholarship to a second year student in an A&P (Aircraft and Powerplant) educational program (two year accredited aviation technical school.)

Eligibility: Applicant must be a third or fourth year student in a four year program at an accredited institution to be considered for the BS in Aviation Management or Professional Pilot programs. Applicant must be a second year student in an A&P program at a two year accredited aviation technical school. Applicant must possess a minimum 3.0 grade point average (GPA). Applicant must submit two references as well as an approximately 500 word essay describing their desire to pursue a career in the aviation field. Applicant must exhibit financial need for consideration.

Award(s) & Amount(s): 2, $1000

❒ AIR TRAFFIC CONTROL ASSOCIATION SCHOLARSHIP

General Counsel
Air Traffic Control Association Scholarship Fund, Inc.
2300 Clarendon Boulevard, Suite 711
Arlington, VA 22201
Tel. (703) 522-5717
Fax (703) 527-7251

Purpose: Awarded to promising men and women who are enrolled in programs leading to a Bachelor' degree or higher in aviation related courses of study and to full-time employees engaged in advanced study to improve their skills in an air traffic control or aviation discipline.

Eligibility: Applicants must be either half or full-time students enrolled in an accredited college program and pursuing an aviation-related field. Must be an U.S. citizens. Submit an application form along with a 400 word paper which explains "How My Education Efforts Will Enhance My Potential Contribution to Aviation". Two categories of awards 1) Half to full-time students; and 2) Full-time employees engaged in advanced (not necessarily degree) study.

Award(s) & Amount(s): Approximately 6 awards: 3 for $2,500 each for students, 3 for $1,500 each for full time employees. (Number and amount of scholarship awards vary depending upon qualifications, financial need, and number of outstanding candidates who apply.

❐ AVIATION COUNCIL OF PENNSYLVANIA SCHOLARSHIP
Director
Aviation Council of Pennsylvania
3111 Arcadia Avenue
Allentown, PA 18103
Tel. (215) 797-1133

Deadline: July 31

Eligibility: Scholarships are awarded each year to individuals in the Aviation Maintenance, Aviation Management and Aviation Pilot field.

Note: Applicants must be residents of Pennsylvania but can attend school outside Pennsylvania.

Award(s) & Amount(s): 3, $1,000 (amounts may vary)

❐ BOEING STUDENT RESEARCH AWARD
Travel and Tourism Research Association
Attn. TTRA Awards
10200 West 44th Avenue, Suite 304
Wheat Ridge, CO 80033
Tel. (303) 940-6557
Fax: (303) 422-8894

Deadline: February

Purpose: To recognize and reward outstanding student authors of research papers on the travel industry.

Eligibility: Eligible are undergraduate or graduate students enrolled in university degree programs. They must have recently written or be planning to write a research paper on travel or tourism research, or a marketing paper from the standpoint of travel research. The research may deal with primary or secondary data or with a theory. Selection is based on quality of research, creativity of approach, relationship to travel/tourism, usefulness and applicability, and quality of presentation.

Special features: This competition is sponsored by Boeing Commercial Airplanes.

Limitations: Applications must include 3 copies of the complete paper and 3 copies of a 500 - 1,000-word abstract.

Award(s) & Amount(s): 1 first prize and 1 Merit Award winner each year. The winner receives a cash prize of $1,000, complimentary conference registration and travel, and hotel costs up to $1,000. A Merit Award winner receives $250; 3 Honorable Mention winners receive certificates. Duration: This competition is held annually.

❒ THE FRED A. HOOPER MEMORIAL SCHOLARSHIP
Transportation Clubs International Scholarship
1275 Kamus Drive
Fox Island, WA 98333

Deadline: March 31

Eligibility: Must be enrolled in a Transportation or Traffic Management degree program oriented toward airport and airway traffic.

Award(s) & Amount(s): 1, $1,000.

❒ NATIONAL BUSINESS AVIATION ASSOCIATION
Janelle C. Lacoste, Executive Services Coordinator
Freeport-McMoRan Inc.
P.O. Box 6119
New Orleans, LA 70141
Tel. (504) 582-4322
Fax: (504) 582-4661
E-mail: Janell_Lacoste@fmi.com

Deadline: Contact Sponsor

Description: National Business Aviation Association has increased its scholarship for **schedulers and dispatchers** to $10,000 through assistance by Exxon, Aviat and Signature Flight Support. The scholarships are offered to past or present schedulers or dispatchers.

Award(s) & Amount(s): $10,000

❒ SOUTHEASTERN AIRPORT MANAGERS ASSOCIATION
SAMA Treasurer - Sam McKenzie
P.O. Box 35005
Greensboro, NC 27425
Tel. (910) 665-5600

Deadline: Contact Sponsor

Eligibility: (1) One scholarship is given to an individual at each of the following universities: Embry Riddle Aeronautical University, Auburn University, Louisiana Tech University, Middle Tennessee State University and Delta State University; and (2) Institution must be located in the SAMA region and should have an approved four-year program (a Bachelor's degree in a curriculum directly related to airport management).

Special Notes: Scholarships are handled through Embry-Riddle Aeronautical University as noted on the scholarship application.

Award(s) & Amount(s): 5, $1,500

❒ THE WILFRED M. POST, JR. AVIATION SCHOLARSHIP
Northeast Chapter of the American Association of Airport Executives (AAAE)
The Wilfred M. Post, Jr. Aviation Scholarship, c/o AAAE
4212 King Street
Alexandria, VA 22302
Tel. (703) 824-0500 ext. 26

Deadline: March 1

Eligibility: Must be a junior or senior and working towards a Bachelor's degree in a field of Aviation Management. Must submit an application with necessary documentation.

Award(s) & Amount(s): 4, $1,000

AVIATION MAINTENANCE

❐ ADMA SCHOLARSHIP PROGRAM
Aviation Distributors and Manufacturers Association
Charlotte Keyes
1900 Arch Street
Philadelphia, PA 19103-1498
Tel: (215) 564-3484
Fax: (215) 564-2175 or (215) 963-9784

Deadline: March 15

Description: ADMA will award two $1,000 scholarships to selected students each year. These will be selected among the following: One $1,000 scholarship to a Bachelor of Science candidate with an Aviation Management major with emphasis in any of the following four areas: General Aviation, Airway Science Management, Aviation Maintenance and Airway Science Maintenance Management; or, a Bachelor of Science candidate with a major in Professional Pilot with any of the three following emphasis: General Aviation, Flight Engineer, or Airway Science A/C Systems Management; or, one $1,000 Scholarship to a second year student in an A&P (Aircraft and Powerplant) educational program (two year accredited aviation technical school.)

Eligibility: Applicant must be a third or fourth year student in a four year program at an accredited institution to be considered for the BS in Aviation Management or Professional Pilot programs. Applicant must be a second year student in an A&P program at a two year accredited aviation technical school. Applicant must possess a minimum 3.0 grade point average (GPA). Applicant must submit two references as well as an approximately 500 word essay describing their desire to pursue a career in the aviation field. Applicant must exhibit financial need for consideration.

Award(s) & Amount(s): 2, $1000

❐ AVIATION COUNCIL OF PENNSYLVANIA SCHOLARSHIP
Director
Aviation Council of Pennsylvania
3111 Arcadia Avenue
Allentown, PA 18103
Tel. (215) 797-1133

Deadline: July 31

Eligibility: Scholarships are awarded each year to individuals in the Aviation Maintenance, Aviation Management and Aviation Pilot field.

Note: Applicants must be residents of Pennsylvania but can attend school outside Pennsylvania.

Amount(s) Awarded: 3, $1,000 (amounts may vary)

❐ AVIATION MAINTENANCE EDUCATIONAL FUND

Aviation Maintenance Foundation International
Contact: Richard S. Kost
P.O. Box 2826
Redmond, WA 98073
Tel. (360) 658-5272
Fax: (360) 658-1919
Website: http://www.amfic.com
E-mail: amfic@ix.netcom.com

Deadline: Varies

Eligibility: Full-time students attending an FAA approved Part 147 School (Certified Aviation Maintenance Technician School). A stamped (.52 cents), self-addressed envelope must be sent to AMFI to receive scholarship information.

Award(s) & Amount(s): Varies, $250 and $1000; Not Renewable, but reapplication is permitted.

❒ CAREER QUEST SCHOLARSHIP PROGRAM
Sponsors: Professional Aviation Maintenance Association (PAMA) & Superior Air Parts
Professional Aviation Maintenance Association
Scholarship Department
1200 18th St., NW, Suite 401
Washington, DC 20036-2598
Tel. (202) 206-0545

Deadline: October 15 (fall deadline); February 15 (spring deadline)

Purpose: To provide funding to students pursuing A&P Technician certification through an FAR Part 147 Aviation Maintenance Technician School.

Eligibility: (1) The awards are presented to individuals who are enrolled in a Part 147 Maintenance School; and (2) Students must have completed 25 percent of the required program and have maintained at least a "B" average. Students must show a financial need.

Special Notes: (1) Checks from PAMA are written by the National Treasurer to the Chapter, school, or person who has been approved to receive the money; (2) Schools are asked to submit their five best applicants; therefore, individuals must apply through their school; (3) Letters of recommendation must be typed and signed; and (4) The application must be signed by the sponsoring chapter member or institution instructor.

Award(s) & Amount(s): Several, $1,000

❒ THE JOSEPH FRASCA EXCELLENCE IN AVIATION SCHOLARSHIP
The Frasca Family and the University Aviation Association (UAA)
Dr. David A. NewMyer
c/o College of Technical Careers
Southern Illinois University at Carbondale
Carbondale, IL 62901-6621

Deadline: April 15

Eligibility: Must show evidence of EXCELLENCE in activities, studies, events, organizations, etc., related to aviation. Minimum of a 3.0 on a 4.0 scale overall grade point average. Federal Aviation Administration certification/qualifications in either Aviation Maintenance or Flight. Aviation organization memberships such as Alpha Eta Rho, NIFA Flying Team, Experimental Aircraft Association, Warbirds of America, etc. Aviation activities, projects, events, etc., which will demonstrate an interest and an enthusiasm for aviation and must be a junior or senior currently enrolled in a UAA member institution.

Award(s) & Amount(s): 2 pilots and 2 mechanics, $500 scholarships awarded each year at the National Intercollegiate Flying Association National Flying Championships banquet.

❏ **MECHANIC/TECHNICIAN SCHOLARSHIP AWARD PROGRAM**
Helicopter Association International
Tel. (703) 683-4646
Email: scott.dibiasio@rotor.com

Deadline: September

Description: The Helicopter Association International (HAI) is pleased to announce the Mechanic/Technician Scholarship Award Program. This year's award winners will have a choice of one of the following courses made available by the helicopter and engine manufacturers. In addition to tuition, each of four winners will receive $1500, $1000, $750, or $500 respectively, to defer expenses associated with attendance at the course.

Eligibility: Successful applicants will have completed an aviation maintenance program and passed all oral, practical, and FAA written exams.

Award(s) & Amount(s): Varies

❏ **PAMA STUDENT SCHOLARSHIP PROGRAM**
Professional Aviation Maintenance Association
636 Eye St., NW, Suite 300
Washington, DC 20001-3736
Tel. (202) 216-9220
Fax: (202) 216-9224

Deadline: Fall applications must be submitted September 1 through October 15; Spring applications from January 1 through February 15.

Purpose: Funds are to be used specifically for tuition, books, and/or other expenses directly related to the A&P program.

Award(s) & Amount(s): Varies

❏ **R. MINKIN AEROSPACE ENGINEERING SCHOLARSHIP**
Contact: Donna Rosenthal
Tel. (612) 625-8000

Description: The Aerospace Engineering and Mechanics Award Committee reviews each eligible undergraduates GPA and makes their selection annually.

Eligibility: Open to a student majoring in Aerospace Engineering and Mechanics. It is awarded to a sophomore student with the best GPA after five consecutive quarters, with at least 72 credits completed satisfactorily.

Award(s) & Amount(s): $800.

❏ **WILLIAM M. FANNING MAINTENANCE SCHOLARSHIP**
National Business Aviation Association
Contact: Annie Brown
1200 Eighteenth St., NW, Suite 400
Washington, DC 20036
Tel. (202) 783-9000
Fax: (202) 331-8364
Website: http://www.nbaa.org

Deadline: Contact Sponsor

Description: The National Business Aviation Association annually award scholarships to applicants who are pursuing careers as maintenance technicians. One award will benefit a student who currently is enrolled in an accredited airframe and powerplant (A&P) program at an approved Part 147 school. The second award will benefit an individual who currently is not enrolled but has been accepted for enrollment in an A&P program. Awards will be made to U.S. citizens without regard to race, sex, religion or national origin.

Eligibility: At the time of application, the applicant either must be officially enrolled in an accredited A&P program, or have been accepted for enrollment by an approved Part 147 school. Proof of enrollment or acceptance of enrollment must be provided prior to the distribution of the award. An official transcript from the applicant's school must be submitted with the scholarship application form (for schools currently enrolled). The completed application must be accompanied by a 250-word, typed, double-spaced essay describing the applicant's interest and goals for a career in the aviation maintenance field. A letter of recommendation from either a faculty member or other individual who is familiar with the applicant's capabilities. A resume must be submitted.

Award(s) & Amount(s): 2, $2,500

AVIONICS / AIRCRAFT ELECTRONICS

❒ AIRCRAFT ELECTRONICS ASSOCIATION EDUCATIONAL FOUNDATION
4217 South Hocker
P.O. Box 1963
Independence, MO 64055-0963
Tel. (816) 373-6565
Fax: (816) 473-8100
http://www.AEAavnews.org
E-mail: aea@microlink.net

This foundation is awarding 24 scholarships to students attending or planning to attend an accredited school in an avionics or aircraft repair program.

❒ Lowell Gaylor Memorial Scholarship
Eligibility: This scholarship is in memory of Lowell Gaylor, President of Avel Company, Dallas, Texas, who supported general aviation and the AEA for over 25 years. This scholarship is available to anyone who plans to or is attending an accredited school in an avionics or aircraft repair program.

Amount: $1,000

❒ Bud Glover Memorial Scholarship
Eligibility: This scholarship is in memory of Bud Glover, former Vice President of General Aviation Sales for Bendix/King Radio. Bud was a 25 year contributor to our avionics industry. This scholarship is available to anyone who plans to or is attending an accredited school in an avionics or aircraft repair program.

Amount: $1,000

❒ Navair Limited Scholarship
Eligibility: For high school seniors and/or college students who plan to or are attending an accredited college/university in an avionics or aircraft repair program. The educational institution must be located in Canada.

Amount: $1,000

❒ **Northern Airborne Technology Scholarship**
Eligibility: For high school seniors and/or college students who plan to or are attending an accredited college/university in an avionics or aircraft repair program. The educational institution must be located in Canada.

Amount: $1,000

❒ **Castleberry Instruments Scholarship**
Eligibility: This scholarship is available to anyone who plans to or is attending an accredited school in an avionics or aircraft repair program.

Amount: $1,000

❒ **Gulf Coast Avionics Scholarship to Fox Valley Technical College**
Eligibility: This scholarship is available to anyone who plans to or is attending Fox Valley Technical College in Oshkosh, WI, in the avionics program.

Amount: $1,000

❒ **Field Aviation Co., Inc.**
Eligibility: For high school seniors and/or college students who plan to or are attending an accredited college/university in an avionics or aircraft repair program. The educational institution must be located in Canada.

Amount: $1,000

❒ **Monte R. Mitchell Global Scholarship**
Eligibility: Sponsored by Mid-Continent Instruments Co., this scholarship is named in honor of retired AEA President, Monte Mitchell. The scholarship is available to a European student pursuing a degree in aviation maintenance technology, avionics or aircraft repair at an accredited school located in Europe or the United States.

Amount: $1,000

❒ **David Arver Memorial Scholarship**
Eligibility: This scholarship is given by Dutch and Ginger Arver in memory of their son, David. Dutch Arver has been a strong supporter of AEA for many years and served on the Board of Directors until his retirement. The recipient of this scholarship shall be any student who plans to attend an accredited vocational/technical school located in AEA Region III. (Illinois, Indiana, Iowa, Kansas, Michigan, Minnesota, Missouri, Nebraska, North Dakota, South Dakota or Wisconsin. Student must enroll in an avionics or aircraft repair academic program.

Amount: $1,000

❒ **Russell Leroy Jones Memorial Scholarship to Colorado Aero Tech**
Eligibility: The intent of this full tuition scholarship is for attendance at Colorado Aero Tech in Broomfield, Colorado, and covers the Avionics course tuition only. All other costs (tools, fees, room & board) will be the responsibility of the student. This scholarship is open to anyone except students currently enrolled in avionics courses at Colorado Aero Tech.

Amount: 3, $6,000

❒ **Paul and Blanche Wulfsberg Scholarship**
Eligibility: This scholarship is provided by the Paul & Blanche Wulfsberg Foundation. Paul Wulfsberg, founder of Wulfsberg Electronics, has been a supporter of general aviation for many years. This scholarship is available to anyone who plans to or is attending an accredited school in an avionics or aircraft repair program.

Amount: $1,000

❒ **Plane & Pilot Magazine/GARMIN Scholarship**
Eligibility: Open to high school, college or vocational/technical school students who plan to or are attending an accredited vocational/technical school in an avionics or aircraft repair program.

Amount: $2,000

❐ **Terra-By-Trimble Collegiate Scholarship**
Eligibility: Open to high school seniors and college students who are children or grandchildren of employees of AEA regular members and who are planning to attend or are attending an accredited school/college. (Does not have to be an aviation-related field of study.) Award is based on essay competition. Essay shall not be over 2,500 words. Entries must be accompanied by a letter verifying the entrants type of relationship with the AEA regular member. Essay must be double-spaced on 8.5 * 11" white paper with the writer's name at top left corner of each page. Essay will be judged on originality, clarity, writing style, arrangement and definitive treatment of the subject. Essay submission deadline is (contact sponsor). Submit essay to: Terra Corporation, Terra Collegiate Scholarship Award, 3520 Pan American Freeway NE, Albuquerque, NM 87107-4796

Amount: $2,000

❐ **College of Aeronautics Scholarship**
Eligibility: This is a scholarship in our two-year avionics program (Associate in Occupational Studies). It is for the duration of this program (two years). The student selected receives $750 per semester with a maximum of $3,000 (four semesters).

Amount: $3,000

❐ **Leon Harris/Les Nichols Memorial Scholarship to Spartan School of Aeronautics**
Eligibility: This scholarship is to be awarded to an individual desiring to pursue an Associate's Degree in Applied Science in Aviation Electronics (Avionics) at NEC Spartan School of Aeronautics Campus in Tulsa, Oklahoma. The award will cover all tuition expenses for eight quarters or until the recipient completes an Associate's Degree, whichever comes first. All other costs (tools, living expenses and student fees) will be the responsibility of the student. The applicant may not be currently in the Avionics program at Spartan.

Amount: over $16,000

❐ **EDMO Distributors Scholarship**
Eligibility: This scholarship is to be awarded to an individual who holds at least a private pilot certificate and who is seeking a career in aviation. Their main course of studies are to be in avionics at a college or technical school offering an avionics program.

Amount: $1,000

❐ **Robert Kimmerly Memorial Scholarship**
Eligibility: This scholarship is to be awarded to an individual who holds at least a private pilot certificate and who is seeking a career in aviation. Their main course of studies are to be in avionics at a college or technical school offering avionics program.

Amount: $1,000

❐ **Bose Corporation Avionics Scholarship**
Eligibility: This scholarship is available to anyone who plans to or is attending an accredited school in an avionics program.

Amount: $1,500

❐ **Chuck Peacock Honorary Scholarship**
Eligibility: This scholarship is available to all AEA members, their children, grandchildren or dependents. The scholarship may be used for any field of study and will be applied towards their tuition. The scholarship honors the founder of the Aircraft Electronics Association.

Amount: $1,500

❐ **Mid-Continent Instrument Scholarship**
Eligibility: This scholarship is available to anyone who plans to or is attending an accredited school in an avionics program.

Amount: $1,000

❒ **The Jim Cook Honorary Scholarship**
Eligibility: This scholarship is available to all AEA members, their children, grandchildren or dependents. The scholarship may be used for any field of study and will be applied towards their tuition. The scholarship honors the first chairman of the AEA Educational Foundation.

Amount: $1,500

❒ **Gene Baker Memorial Scholarship**
Eligibility: This scholarship is available to all AEA members, their children, grandchildren or dependents. The scholarship may be used for any field of study and will be applied towards their tuition. The scholarship is in memorial to the first Executive Director of the Aircraft Electronics Association.

Amount: $1,500

❒ **Dutch and Ginger Arver Scholarship**
Eligibility: This scholarship is available to anyone who plans to or is attending an accredited school in an avionics program.

Amount: $1,000

❒ **AEA Educational Foundation Pilot Training Scholarship**
Eligibility: This scholarship is available to an avionics technician that is employed by an AEA member. The technician must have been employed for a minimum of one (1) year by the AEA member. The scholarship is to be used to acquire a (private pilot SEL) license. There will be three (3) individual scholarships awarded each year.

Amount: (3) $1,000

❒ **MCCOY AVIONICS SCHOLARSHIPS**
Aircraft Electronics Association (AEA)
10761 Watkins Road
Marysville, OH 43040
Tel. (800) 654-8124

Deadline: March 1

Eligibility: Applicants include high school students who are the children or grandchildren of employees of regular AEA members. These applicants can attend a summer camp specializing in acquainting participants with the field of aviation. This award is not available for use in gaining flight instruction or in attaining flight rating.

Award(s) & Amount(s): 2, $500

❒ **WILLIAM E. JACKSON AWARD**
Radio Technical Commission for Aeronautics
1140 Connecticut Ave., NW, Suite 1020
Washington, DC 20036
Tel. (202) 833-9339
Fax (202) 833-9434

Deadline: June 30

Eligibility: Scholarship is given to an undergraduate or graduate student majoring in aviation electronics or telecommunications.

Award(s) & Amount(s): 1, $2,000

FELLOWSHIPS

❒ AMELIA EARHART FELLOWSHIP AWARDS
Zonta International Foundation
557 W. Randolph Street
Chicago, IL 60661-2206
Tel. (312) 930-5848

Deadline: November 1

Eligibility: Grants in the amount of $6,000 for graduate study in aerospace-related science or engineering are awarded annually and are available to qualified women. Qualifications include a Bachelor's degree which qualifies the candidate for graduate work in aerospace-related fields, acceptance by an institution offering fully accredited graduate courses and degrees in aerospace-related science or engineering, good transcript, and career goals.

Award(s) & Amount(s): Several, $6,000

❒ ARIZONA/NASA SPACE GRANT GRADUATE FELLOWSHIP
The University of Arizona, Arizona State University & NASA Space Grant Program
Lunar & Planetary Laboratory
The University of Arizona
Tucson, AZ 85721
Tel. (520) 621-8556
Fax (520) 621-4933
Website: http://www.seds.org/spacegrant.htm

Contact: Susan A. Brew, Sr. Coordinator
E-mail: sbrew@seds.org

Deadline: Early Spring (review the website for possible changes)

Description: Part- and Full-time graduate fellowships are available at The University of Arizona (outreach-program requirement) and Arizona State University for space and related science and engineering disciplines. Applicants must be U.S. Citizen and must have status of a graduate student.

Award(s) & Amount(s): Various at Arizona State University; 2 year-awards, $16,000 plus tuition, fees, and $1,500 for travel at The University of Arizona

---

❒ FRANCES SHAW FELLOWSHIP/INTERNSHIP
Ragdale Foundation
1260 N. Green Bay Road
Lake Forest, IL 60045
Tel. (708) 234-1063

Deadline: February

Eligibility: This is an annual fellowship given to a woman who began writing serious after the age 55. Fellows are flown to Ragdale from anywhere in the continental U.S. and receive a free 2 month residency.

❒ NATIONAL AIR AND SPACE MUSEUM FELLOWSHIPS
Contact: Collette Williams, Program Coordinator

Office of Fellowship and Grants
Smithsonian Institution
955 L'Enfant Plaza, Suite 7000
Washington, D.C. 20560-0312
Tel. (202) 357-2515
Fax (202) 786-2447

Deadline: January 15

❒ A. Verville Fellowship
Desacription: The A. Verville Fellowship is a competitive nine- to twelve-month in-residence fellowship intended for the analysis of major trends, developments, and accomplishments in the history of aviation or space studies. The fellowship is open to all interested candidates with demonstrated skills in research and writing. An advanced degree in history, engineering, or related fields is not a requirement. Graduate predoctoral students will normally not be considered for the Verville; they should apply for a Guggenheim Fellowship. Residence in the Washington, D.C., metropolitan area during the fellowship term is a requirement of these fellowship.

Award(s) & Amount(s): A stipend of $35,000 will be awarded for a 12-month fellowship, with limited additional funds for travel and miscellaneous expenses.

❒ Guggenheim Fellowship
Description: The Guggenheim Fellowship is a competitive three to twelve-month in-residence fellowship for pre- or postdoctoral research. Scholars interested in historical research related to aviation and space are encouraged to apply. Predoctoral applicants should have completed preliminary course work and examinations and be engaged in dissertation research. Postdoctoral applicants should have received their Ph.D. within the past seven years. All applicants must be able to speak and write fluently in English. Residence in the Washington, D.C., metropolitan area during the fellowship term is a requirement of the fellowship.

Award(s) & Amount(s): A stipend of $15,000 for predoctoral candidates and $27,000 for postdoctoral candidates will be awarded, with limited additional funds for travel and miscellaneous expenses.

❒ Ramsey Fellowship In Naval Aviation History
Description: The Ramsey Fellowship in Naval Aviation History is a competitive nine-to-twelve month, in-residence fellowship focused on the history of aviation at sea and in naval service, particularly in the U.S. Navy. The fellowship is open to all interested candidates with demonstrated skills in research and writing. An advanced degree is not a requirement. Residence in the Washington, D.C., metropolitan area during the fellowship term is a requirement of these Fellowships.

Award(s) & Amount(s): A stipend of $40,000 will be awarded for a 12-month fellowship, with limited additional funds for travel and miscellaneous expenses.

❒ NEBRASKA SPACE GRANT SCHOLARSHIPS AND FELLOWSHIPS
Nebraska Space Grant Consortium
UNO Aviation Institute
Allwine Hall 422
University of Nebraska at Omaha
Omaha, NE 68182-0508
Tel. (402) 554-3772
Fax: (402) 554-3781
E-mail: nasa@cwis.unomaha.edu

Purpose: To fund aerospace-related research and studies on the undergraduate and graduate school level for students in Nebraska.

Eligibility: This program is open to all eligible undergraduate and graduate students at the following schools in Nebraska: University of Nebraska at Omaha, University of Nebraska at Lincoln, University of Nebraska at

Kearney, University of Nebraska Medical Center, Creighton University, Western Nebraska Community College, Chadron State College, and Nebraska Indian Community College. Applicants must be U.S. citizens and working on a degree in an aerospace-related area. Special attention is given to applications submitted by women, underrepresented minorities, and individuals with disabilities.

Special features: Recipients conduct research in an aerospace-related area and receive at least 3 semester credits for that activity during the year of the award. Funding for this program is provided by the National Aeronautics and Space Administration.

Limitations: Recipients must submit a progress report each semester on the aerospace project to their designated faculty monitor. Failure to provide that report disqualifies the student from reapplying for a renewal fellowship.

Duration: Academic awards are 1 year; summer awards are for the summer months. Both awards are renewable.

Award(s) & Amount(s): Academic year awards are $7,500; summer awards are $2,500.

❒ TENNESSEE SPACE GRANT CONSORTIUM GRADUATE FELLOWSHIP PROGRAM
Tennessee Space Grant Consortium
Vanderbilt University; Box 1592, Station B
Nashville, TN 37235
Tel. (615) 343-1148
Fax (615) 343-6687
Website: http://vuse.vanderbilt.edu/~tnsg/homepage.html
E-mail: eweiss@vuse.vanderbilt.edu

Contact: Alvin M. Strauss, Director; Ellie Weiss Rosenbloom, Program Coordinator

Deadline: Varies

Description: The Tennessee Space Grant Consortium is a grant from NASA to promote space and science education from K to 12th grade through the graduate level. The applicant must be a U.S. citizen. The fellowship must be used for research.

Award(s) & Amount(s): Varies

❒ TRANSPORTATION RESEARCH BOARD GRADUATE RESEARCH AWARD PROGRAM VIII
Smithsonian Institute Transportation Research Board
2101 Constitution Ave, NW
Washington, D.C.
Tel. (202) 334-3206
Fax (202) 334-2003

Award(s) & Amount(s): A stipend of $6,000 paid in progress payments of 25 percent installments during the research, with final payment on completion and acceptance of the paper.

❒ U.S. AIR FORCE DISSERTATION YEAR FELLOWSHIP IN U.S.
Military Aerospace History
Attn. Major Myrtistene H. Wilson
Office of Air Force History
Headquarters USAF (AF/HO) Bldg. 5681
Bolling AFB
Washington, D.C. 20332-6098
Tel. (202) 767-5088
Fax (202) 404-7915

Deadline: March 13

Award(s) & Amount(s): Two fellowships with stipends of $10,000 each are awarded each year.

FLIGHT TRAINING

❐ ADMA SCHOLARSHIP PROGRAM

Aviation Distributors and Manufacturers Association
Charlotte Keyes
1900 Arch Street
Philadelphia, PA 19103-1498
Tel: (215) 564-3484
Fax: (215) 564-2175 or (215) 963-9784

Deadline: March 15

Description: ADMA will award two $1,000 scholarships to selected students each year. These will be selected among the following: One $1,000 scholarship to a Bachelor of Science candidate with an Aviation Management major with emphasis in any of the following four areas: General Aviation, Airway Science Management, Aviation Maintenance and Airway Science Maintenance Management; or, a Bachelor of Science candidate with a major in Professional Pilot with any of the three following emphasis: General Aviation, Flight Engineer, or Airway Science A/C Systems Management; or, one $1,000 Scholarship to a second year student in an A&P (Aircraft and Powerplant) educational program (two year accredited aviation technical school.)

Eligibility: Applicant must be a third or fourth year student in a four year program at an accredited institution to be considered for the BS in Aviation Management or Professional Pilot programs. Applicant must be a second year student in an A&P program at a two year accredited aviation technical school. Applicant must possess a minimum 3.0 grade point average (GPA). Applicant must submit two references as well as an approximately 500 word essay describing their desire to pursue a career in the aviation field. Applicant must exhibit financial need for consideration.

Award(s) & Amount(s): 2, $1000

❐ THE AERO CLUB OF NEW ENGLAND

A.C.O.N.E. Education Committee
Contact: Patti Sanford
4 Emerson Drive
Action, MA 01720
Tel. (978) 263-7793

Description: The Aero Club of New England, in an effort to assist and promote professional aviation careers, has over the past ten years award scholarships to men and women totaliling more than $85,000. Some of these scholarships are matching fund scholarships wherein the student and the Aero Club of New England jointly pay for the trainig in am matching process up to the limit of the scholarship.

Eligibility: The appicant must intend to pursue a professional aviation career; hold a current Airmen and Medical certificate; have accumualated 160 hours of total flight time (100 hours for the Ehlers Scholarship); **Be a New England Resident; attend an FAA approved Flight School in New England;** use the scholarship funds in one year; be at least 18 years of age; demonstrate a financial need for the award; and have a current Biennial Flight Review.

❐ New Advanced Pilot Scholarships

Description: Advanced Pilot Scholarships have been increased to seven in number, providing $10,500 in flight training. These memorial scholarships are worth $1,000 scholarships for flight training: The Bauer-Bisgeier Memorial Scholarship is provided each year; The Andrew Channing Cabot Memorial Scholarship is provided each year; The Edward D. Waters Memorial Scholarship; The Florence M. Abely Memorial Scholarship.

❒ Ongoing Advanced Pilot Scholarships

- The Ann Wood-Kelly Scholarship & the ACONE Honored Member Scholarship, each $2,000
- The Charles and Arlene Ehlers Memorial Scholarship, $1,500 for instrument training
- The Bonita Connors Memorial Scholarship for $500.

Award(s) & Amount(s): Varies

❒ AMELIA EARHART MEMORIAL SCHOLARSHIP
Ninety-Nines, Inc.
International Women Pilots
Will Rogers World Airport
P.O. Box 59965
Oklahoma City, OK 73159
Tel. (405) 685-7969

Purpose: To provide financial support to members of the Ninety-Nines who are interested in advanced flight training courses in aviation.

Eligibility: Eligible to apply are Ninety-Nine members who wish to take advanced flight training or courses in specialized branches of aviation.

Award(s) & Amount(s): 10 or more each year. Up to $1,000 per year. Duration: 1 year.

❒ AMERICAN AIRLINES FIRST OFFICER CANDIDATE COURSE SCHOLARSHIP
American Airlines
Commercial Flight Training
P. O. Box 619617
MD 821, GSWFA,
DFW Airport, TX 75261-9617
Tel. (800) 678-8686

Deadline: May 1

Purpose: To provide a "bridge" between university aviation flight education and the airline cockpit and to promote interest in the First Officer Candidate Course. Applicants must have, or be about to receive, a four year degree from a University Aviation Association member institution.

Eligibility: Applicants must be within one semester of graduation or no more than one year beyond graduation at the time of the deadline. Applicants must be 21 years of age by the application deadline. Applicants must have a 3.5 overall GPA on a 4.0 scale. Applicants must hold the FAA commercial certificate with instrument and multi-engine ratings. Applicants must have 400 total flight hours and at least 40 hours of multi-engine time. Applicants must have completed their university's FAA-approved flight program. There are more criteria. Applications are available from UAA member institutions as well as from American Airlines.

Award(s) & Amount(s): 2 awards for $8,500 each to be used for the American Airlines First Officer Candidate Course at Eagle Training Center.

❒ AMERICAN HELICOPTER SOCIETY
217 N. Washington Street,
Alexandria, VA 22314
Tel. (703) 684-6777

Description: Awards for undergraduate and graduate students pursuing studies in vertical flight engineering.

Award(s) & Amount(s): Range between $1,000 - $2,000

❒ ANNE MARIE MORRISSEY AVIATION SCHOLARSHIP

106 NW Hackberry
St., Lee's Summit, MO 64064-1435

Deadline: January 1

Description: The Anne Marie Morrissey Aviation Scholarship fund is for flight training. Applicants may apply for training in any aviation discipline provided the training is completed within eight months after receiving the award. Applicants must have a valid medical certificate appropriate for the rating or license being sought, and also have the initial qualification for that rating. The scholarship will only apply to the flight portion of the training, and will cover 75% of the cost of each lesson.

Award(s) & Amount(s): One or more; $6,500

❐ AVIATION COUNCIL OF PENNSYLVANIA SCHOLARSHIP
Director
Aviation Council of Pennsylvania
3111 Arcadia Avenue
Allentown, PA 18103
Tel. (215) 797-1133

Deadline: July 31

Eligibility: Scholarships are awarded each year to individuals in the Aviation Maintenance, Aviation Management and Aviation Pilot field.

Note: Applicants must be residents of Pennsylvania but can attend school outside Pennsylvania.

Amount(s) Awarded: 3, $1,000 (amounts may vary)

❐ CADET YOUTH FLIGHT SCHOLARSHIP
Soaring Society of America, Inc.
PO Box E
Hobbs, NM 88241
Tel. (505) 392-1177

Deadline: June 30

Purpose: To finance study of sailplane flying at any USA training operation, commercial or club-based, for hobby as well as career goals. The program may be of special interest to prospective Aviation or Aerospace Engineering students.

Eligibility: Ages 14-22 inclusive. Not holder of any FAA pilot license. Interest in soaring or other form of flying. Criteria are: 1) Original Essay on "Some aspect of soaring flight" (600 words); 2) Desire to fly; 3) Involvement in aviation activities; 4) Outstanding achievement in some non-aviation area (academic, sports, social); 5) Financial need sufficient to preclude flying lessons.

Award(s) & Amount(s): Up to 2 scholarships annually, 3 second prizes, 5 third prizes. $600 for sailplane flying lessons. For training fees and lesson costs only (First Prize). Second prize: Set of soaring textbooks. Third prize: Youth membership in SSA.

Special Instructions: Must be made on a form available ONLY at active soaring sites. Distribution to these sites is made in early April. There are over 250 soaring Clubs or Schools in the USA. A full list of locations may be obtained from SSA: PO Box E, Hobbs, NM 88241. (505)392-1177.

❐ COMAIR AIRLINE PILOT SCHOLARSHIP
Comair Aviation Academy
Contact: Susan Burrell
2700 Flight Line Ave
Sanford, FL 32773
Tel. (800) 822-5963

Fax: (407) 323-3817
Website: http://www.comairacademy.com
E-mail: comairacademy@msn.com

Deadline: July

Description: Flight training for commercial airline career.

Eligibility: Write an essay, complete an application, desire, motivation

Award(s) & Amount(s): $1,000

❐ DAN L. MEISINGER, SR. MEMORIAL LEARN TO FLY SCHOLARSHIP
National Air Transportation Foundation
4226 King St.
Alexandria, VA 22302
Tel. (703) 845-9000

Deadline: November

Eligibility: This scholarship is to be used for the express purpose of the initial or primary flight training. Must be a college student currently enrolled in an aviation program; Be a high academic achiever, with "B" or better average; Be recommended by an aviation professional, although they may apply directly themselves; and Preferably be a resident of Kansas, Missouri or Illinois.

Awards(s) & Amount(s): $2,500

❐ EAA AVIATION ACHIEVEMENT SCHOLARSHIPS
Experimental Aircraft Association Foundation (EAA)
Scholarship Program
P. O. Box 3065
Oshkosh, WI 54903-3065
Tel. (414) 426-4888

Deadlines: April 1 (Depend on funding availability)

Purpose: Presented by the EAA to individuals active in sport aviation endeavors to further their aviation education or training. Scholarships are not based on endowment and are sometimes not awarded.

Award(s) & Amount(s): 2, $300

❐ FLORENZA DE BANARDI MERIT AWARD
ISA International Career Scholarship & International Society of Women Airline Pilots
International Society of Women Airline Pilots
P.O. Box 38644
Denver, CO 80238

Deadline: January 15

Eligibility: Applicants must have a minimum of 750 total flight hours, a Commercial Pilot Certificate, and a first class medical certificate. Awards will be based on financial need.

Award(s) & Amount(s): Contact Sponsor

❐ FRANK P. LAHM, FLIGHT 9 SCHOLARSHIPS
The Order of Daedalians
P.O. Box 249
Randolph AFB, TX 78148

Deadline: February

Description: This National program presents awards to distinguished aviation organizations and individuals, and awards medals and certificates to ROTC, Civil Air Patrol, and other aviation oriented organizations. Individuals do not apply for these awards; rather, the National organization seeks out organizations and individuals for suitable candidates. Frank P. Lahm Flight 9 participation is generally limited to presenting National awards to close-by organizations and individuals.

An ongoing fund provides scholarships to outstanding junior year college ROTC cadets from local area universities and to outstanding high school senior ROTC cadets from local area schools. College applicants must be planning to attend pilot training after commissioning, and high school applicants must enroll in college ROTC the autumn following their selection. The Flight 9 program varies from year to year, dependent on available funds and military requirements.

Qualifications: College applicants are enrolled in their junior year, and must be qualified to attend pilot training. High school applicants are in their senior year, and applicants must, if selected, enroll in college ROTC the subsequent autumn to receive their awards.

Award(s) & Amount(s): Varies

❒ GOGOS SCHOLARSHIPS PROGRAM
Collegiate Soaring Association, Inc. (CSA)
CSA Gogos Scholars
4671 Kipling #68
Wheat Ridge, CO 80033

Deadline: April 30

Description: For a limited time, youth grants for soaring study. These will be adminstered under the contract by a full-service FBO, will enable in stages the full range of flight training in sailplanes from first flight to advanced FAI badges, and will cover all on-site costs including flight charges, books, room and board, and local transportation.

CSA has scheduled 4 grant awards as a test of the process, and plans to awards 7 grants of $2,000 each in 1999, 2000, and 2001, at which point the funds will be exhausted. Grants will be targeted towards training in the following categories: A (first flight to solo), C (solo to FAA private license), Silver (XC to FAI badgers), and Diamond (advanced badgers, wave , racing). Of the 7 yearly grants, 2 will be earmarked for college students active in CSA and 2 more may be offered through youth partners of the SSA such as Civil Air Patrol and Exploring.

Eligibility: Applicant must be a US Citizen or permanent resident; Student (full-time at an academic institution within one year of grant); Age limit: 14 - 25 (inclusive)

Award(s) & Amount(s): Varies

❒ JAMES R. MIRESSE SCHOLARSHIP
Contact: James R. Miresse
4 Okedee
Hilton Head, SC 29428

Deadline: Contact Sponsor

Description: Half of any major course up to $16,000

Eligibility: Must receive a college degree, aviation oriented with a varied background. Must work 3 to 5 years a job after college.

Award(s) & Amount(s): $16,000

❒ JOHN E. GODWIN, JR. MEMORIAL SCHOLARSHIP FUND

National Air Transportation Foundation
4226 King St.
Alexandria, VA 22302
Tel. (703) 845-9000

Deadline: November

Eligibility: This scholarship is for flight training or rating. Must be at least 18 years of age or older, possess a student pilot certificate with a 3rd class medical certificate and can qualify for a second class medical certificate. Must demonstrate commitment to General Aviation. Be nominated and endorsed by a representative of a Regular or Associated Member Company of the national Air Transportation Association. While not a prerequisite, membership in good standing in the civil air patrol will be favorably considered.

Note: The award is for one year only, can be renewable for one year if a certain academic performance rating is attained. Students must: 1) be in active pursuit of training for their license and/or rating, 2) accumulate a minimum of fifteen dual or solo flight hours each calendar month; and 3) receive a log book endorsement from the Chief Flight Instructor certifying satisfactory achievement for each training stage check.

Award(s) & Amount(s): $2500

❒ THE JOSEPH FRASCA EXCELLENCE IN AVIATION SCHOLARSHIP
The Frasca Family and the University Aviation Association (UAA)
Dr. David A. NewMyer
c/o College of Technical Careers
Southern Illinois University at Carbondale
Carbondale, IL 62901-6621

Deadline: April 15

Eligibility: Must show evidence of EXCELLENCE in activities, studies, events, organizations, etc., related to aviation. Minimum of a 3.0 on a 4.0 scale overall grade point average. Federal Aviation Administration certification/qualifications in either Aviation Maintenance or Flight. Aviation organization memberships such as Alpha Eta Rho, NIFA Flying Team, Experimental Aircraft Association, Warbirds of America, etc. Aviation activities, projects, events, etc., which will demonstrate an interest and an enthusiasm for aviation and must be a junior or senior currently enrolled in a UAA member institution.

Award(s) & Amount(s): 2 pilots and 2 mechanics, $500 scholarships awarded each year at the National Intercollegiate Flying Association National Flying Championships banquet.

❒ Lt. KARA HULTGREEN MEMORIAL SCHOLARSHIP
Women Military Aviators, Incorporated
Major Diana L. Davis
23315 Joy St.
St. Clair Shores, MI 48082

Deadline: May

Eligibility: Interested applicants should complete an application which includes general information, financial information and an essay. Flight instructor or flight school must fill out a financial information form. Members or relatives of members of the scholarship selection committee are not eligible for this award.

Award(s) & Amount(s): 1, $2,500

❒ MARY J. MCGRATH SCHOLARSHIP
Vermont Aviation Advisory Council
PO Box 1527
Montpelier, VT 05601-1527

Deadline: Contact Sponsor

Eligibility: **For a Vermont pilot pursuing their CFI.**

Award(s) & Amount(s): $1,000

❐ MINNESOTA AVIATION TRADE ASSOCIATION FLIGHT SCHOLARSHIP
Sherman Booen
PO Box 23164
Minneapolis, MN 55423
Tel. (612) 869-7026

Purpose: The flight training will be provided by a MATA-member approved flight school.

Award(s) & Amount(s): 3 awarded for $1,000, $750 and $500

❐ REVOLUTION HELICOPTER CORPORATION
1905 W. Jesse James Rd.
Excelsior Springs, MO 64024
Tel: 816-637-2800
Fax: 816-637-7936
Website: http://www.revolutionhelicopter.com
E-mail: sales@revolutionhelicopter.com

Deadline: Contact Sponsor

Description: Revolution Helicopter Corp., manufacturer of the Mini-500 kit-built helicopter, is offering scholarships to Mini-500 customers.

Award(s) & Amount(s): $1,000

❐ THE RICHARD LEE VERNON AVIATION SCHOLARSHIP
Contact: EAA Aviation Foundation
PO Box 3065
Oshkosh, WI 54903-3065

Purpose: To award an individual seeking improved aviation skills for recreation or a career.

Award(s) & Amount(s): 1, $400

❐ SIMUFLITE ADVANCED FLIGHT CREW TRAINING SCHOLARSHIP
SimuFlite, Incorporated
College of Technical Careers
Attn. SimuFlite Scholarship
Southern Illinois University at Carbondale
Carbondale, IL 62901-6621
Tel. (618) 453-8898

Simuflite Training International
Contact: Karen Montalvo
P.O. Box 619119
Dallas/Ft. Worth Airport, Texas 75261
Tel. (800) 527-2463 x8113

Deadline: April 1

Eligibility: This scholarship is for current students of, or graduates of, University Aviation Association (UAA) member institutions. Current students must be graduating by May of the year of application, or must not have graduated more than one year before the application deadline. Applicants must be candidates for, or have received, an aviation-related Baccalaureate degree from a UAA member institution. Must have a minimum of 3.25 (on a 4.0 scale) overall cumulative GPA. Must possess a current first class medical, Commercial Pilot Certificate with Multi-engine and Instrument rating, and a Certified Flight Instructor rating with Instrument Airplane endorsement. Must have flown a minimum of 50 hours PIC or SIC within the previous 12 months. Must be able to schedule training 30 days in advance and demonstrate an interest in corporate or business aviation by writing a 250-word essay on how this scholarship will fit into their plans to enter corporate/business aviation, or aviation in general.

Award(s) & Amount(s): 4, $10,500 corporate aircraft training scholarships

---

❒ SOARING SOCIETY OF AMERICA YOUTH SOARING SCHOLARSHIP
Cadet Scholarship
PO Box E
Hobbs, NM 88241
Tel. (505) 392-1177

Eligibility: Sailplane flying lessons, and lesser prizes of textbooks and memberships. Young persons between 14 - 22 and not holders of any FAA pilot license.

Award(s) & Amount(s): $600

❒ SPORTY'S AVIATION SCHOLARSHIP PROGRAM
Sporty's Pilot Shop
PO Box 44327
Cincinnati, OH 45244
Tel. (513) 735-9000
Fax (513) 735-9200

Deadline: Applications postmarked by January 15, 1997

Purpose: The funds may be used for flight training expenses leading to a recreational, Private, Commercial, or Flight Instructor Certificate, and may include Instrument or Multi-Engine ratings. Eligible expenses include aircraft rental, flight and ground instruction, and the purchase of aviation related educational materials and related pilot supplies.

Eligibility: An applicant must be full-time high school or college student. The student may be pursuing any course of study he or she chooses.

Award(s) & Amount(s): 2 scholarships, $15,000 each and are to be used over a two-year period for pilot training purposes

---

❒ US AIR FORCE ROTC (3 AND 4 YEAR SCHOLARSHIP PROGRAM)
Tel. (612) 962-6320

Eligibility: To be used at any campus which offers Air Force ROTC.

Award(s) & Amount(s): Over $,4000 scholarships awarded to high school seniors per year

---

❒ VIRGINIA AIRPORT OPERATORS COUNCIL AVIATION SCHOLARSHIP AWARD
VOAC Scholarship
Attn. Betty Wilson
5701 Huntsman Road
Sandston, VA 23150
Tel. (800) 292-1034

Deadline: February 16

Eligibility: High school Senior with a "B" average or better planning a career in aviation who has been accepted and is enrolled in an credited college in an aviation-related program. **Limited to Virginia residents.**

Award(s) & Amount(s): 1, $2,000

❒ **THE WAGNER FOUNDATION PROFESSIONAL PILOT SCHOLARSHIP**
EAA Aviation Foundation
PO Box 3065
Oshkosh, WI 54903-3065

Eligibility: An individual pursuing a career as a professional pilot

Award(s) & Amount(s): 1, $200

Grants for Educators & Chapters

❒ **AEROSPACE EDUCATION FOUNDATION**
1501 Lee Highway
Arlington, VA 22209-1198

❒ **Direct Grants to Educators**
Description: AEF offers Educator Grants to promote aerospace education for teachers where funds are not otherwise available. Educator Grants provide up to $250 per academic year in support to elementary and secondary classrooms for aerospace education programs, opportunities, and activities. Fax reply service 1-800-232-3563 document #840

Use of Grants: Funds may be used for any aerospace related activity, from purchasing textbooks or videotapes, to going on a field trip to an aerospace museum, Air Force base or other aerospace facility. The funds may be divided between several aerospace activities.

Note: The application will ask for such information as: Sponsoring teacher's name, school address and subject and grade. Purpose of the grant and amount requested. Number & age of students benefiting from the grant. Previous aerospace related activities in the classroom. Verification/certification by Principal/Instructor.

❒ **Direct Grants to AFJROTC Units and Civil Air Patrol Squadrons**
Description: AEF offers Grants to AFJROTC Instructors and CAP squadrons to promote aerospace education for instructors where funds are not otherwise available. The Grants provide up to $250 per calendar year in support to classroom aerospace educational programs, opportunities, and activities. AEF hopes instructors will find ways to enhance students' ideas on how aerospace plays a prominent role in tomorrow's world. Fax reply service 1-800-232-3563, document #854 for CAP, document #853 for AFJROTC

Use of Grants: Funds can be used for anything related to aerospace, from purchasing textbooks or videotapes, to going on a field trip to an aerospace museum, Air Force base or other aerospace facility. The Grant may not be used for purchasing uniforms, and honor guard or color guard activities.

Note: The application will ask for such information as: Sponsoring instructor's name, School name and address/phone number. Subject and grade; Purpose of the grant and amount requested; Number & age of students to benefit from the grant. Previous aerospace related activities in the classroom. Verification/certification by Principal and the Instructor.

❒ **Chapter Matching Grants for Aerospace Education Programs**
Description: The Chapter Matching Grant program was established to promote aerospace education activities at chapter level where funds are limited. It gives chapters an opportunity to enhance their community's ideas about how aerospace plays a prominent role in today and tomorrow's society. For each dollar a chapter contributes to an aerospace education program, the chapter may request up to $1,000 in matching funds.

Note: Typically chapters request Matching Grants for: Science Fairs, Field Trips, Aerospace activities with schools such as JROTC units, and Career Days.

Matching Grants May Not Be Used For: Scholarships, Non-math/science/technology/aerospace activities, and To pay for the chapter's share of Visions of Exploration.

Applications for matching grants are evaluated by AEF based on the importance and impact of the chapter's activity. Within four to six weeks, AEF will let your chapter contact person know if the grant was approved or not. Chapters may apply for more than one grant per calendar year as long as the total amount requested is less than $1,000.

When a chapter receives a grant, the check will sent to the chapter contact. AEF would like to hear how the activity turned out. Please send photos (action shots) and a short description on how the activity went.

Application: 1. Chapter Name, 2. Chapter contact (any chapter member) with phone and address, 3. Description of activity, location, date and number of people involved/benefiting, 4. Budget for the activity/project, 5. Amount requested (up to $1,000) and amount being contributed by the chapter, 6. Signatures of chapter president and treasurer.

MILITARY AFFILIATION

❒ **AEROSPACE EDUCATION FOUNDATION**
1501 Lee Highway
Arlington, VA 22209-1198
Tel. (800) 727-3337 or (703) 247-5839

Many of these organizations require a military affiliation and provide assistance only for aviation and/or science related education studies. The accuracy of all scholarship and grant information should be verified with each organization.

❒ **AEF Air Force Spouse Scholarships**
Eligibility: Scholarships are awarded to spouses of Air Force members. The spouse can be pursuing an undergraduate or graduate degree. Scholarship recipients are determined by the selection criteria and eligibility listed below. The AEF Scholarships for Air Force Spouses are made possible through contributions from Air Force Association members and chapters.

Scholarships are given for undergraduate or graduate study prior to the Spring semester. Scholarships may be used to pay for any reasonable cost related to pursuing a degree and checks are sent directly to recipient schools.

Award(s) & Amount(s): 30, $1,000

❒ **Aerospace Education Foundation Air Force Spouse Scholarship**
Eligibility: Scholarships are awarded internationally to spouses of Air Force active duty, Air National Guard and Air Force Reserve members during the Spring semester. Applicants must be pursuing their undergraduate, graduate or post-graduate degrees. Spouses who are military members are not eligible. Applications will be available in August at Base Education Offices, Family Support Centers or on the AEF Fax Reply service at 1-800-232-3563. Contact Tel. 1-800-727-3337 ext. 4880

Award(s) & Amount(s): 30, $1,000

❒ **AFROTC Angel Flight and Silver Wings Scholarships** Deadline: January
Description: Aerospace Education Foundation's Angel Flight/Silver Wings Scholarship program provides grants to rising juniors and seniors who are active members of Angel Flight/Silver Wings.

This award, started in 1993, was established to recognize members of AnF/SW for their academic excellence, their continued support of their communities and their dedication to the goals and initiatives of ROTC and the United States Air Force. Angel Flight Scholarships are to be used for the purpose of tuition assistance. Funding for the Angel Flight/Silver Wings Scholarships is provided by AFA Members and Chapters worldwide.

Eligibility: Active member of Angel Flight/Silver Wings for a minimum of one year. Junior or Senior student for academic year. Minimum Grade Point Average (GPA) of 3.0. Ability to demonstrate active involvement in campus and community service projects. Projects positive image of AnF/SW, AFROTC, USAF.
Recommended by an AF/SW advisor.

Award(s) & Amount(s): $1,000

❐ Dr. Theodore von Kármán Graduate Scholarship Program
Description: Scholarships awarded annually to graduating Air Force ROTC seniors pursuing graduate degrees in science, math or engineering.

Award(s) & Amount(s): $5,000

❐ Eagle Grants Enlisted Tuition Assistance
Eligibility: Grants to Community College of the Air Force (CCAF) graduates awarded twice yearly (October & April) as acknowledgment of educational achievement, and to those who intend to pursue a bachelor's degree. This valuable program assists over 500 Air Force enlisted people each year.

Award(s) & Amount(s): $250

❐ Janet R. (Wisemandle) Whittle Memorial Scholarship
Description: In 1997, the Aerospace Education Foundation established a "Janet R. (Wisemandle) Whittle Memorial Scholarship," in memory of Janet Whittle, an Air Force spouse who passed away 1996. Whittle requested that part of her estate be donated to fund a scholarship for spouses of Air Force personnel holding the rank of E-4 or below.

Applications will be available in August at base education offices, family support centers or by calling the Aerospace Education Foundation's fax reply service.

Note: Proof of acceptance into an accredited undergraduate or graduate degree program. Three letters of recommendation. Verification of spouse status from CBPO, first sergeant or commander. Minimum GPA of 3.0 in college; a high school GPA is only accepted if the current semester is the individual first college semester. Two page, double spaced, description of academic goals and community service.

Eligibility: Spouses of Air Force active duty, Air National Guard or Air Force Reserve members. Spouses who are themselves military members are not eligible.

Award(s) & Amount(s): $500

❐ AIR FORCE AID SOCIETY
1745 Jeff Davis Hwy, #202
Arlington, VA 22202-3410
Tel. (800) 429-9475 or (703) 607-3072
DSN 327-3072AFAS h

Eligibility: Applicants include dependent sons and daughters of active duty, retired and deceased Air Force members for undergraduate studies. Spouses of Active Duty members stationed stateside who meet eligibility requirements are now eligible to participate.

Award(s) & Amount(s): $1,000

❒ AIR FORCE SERGEANTS ASSOCIATION (AFSA) AND AIRMEN MEMORIAL FOUNDATION (AMF) SCHOLARSHIPS
AFSA/AMF Scholarship Programs
P.O. Box 50
Temple Hills, MD 20757-0050
Tel. (800) 638-0594

Deadline: April

Description: The Air Force Sergeant Association (AFSA) and the Airmen Memorial Foundation (AMF) join together to financially assist the undergraduate studies of single eligible dependent children of Air Force enlisted members on active duty, serving in the ANG/AFRC, or retired from any of these components. Scholarships are awarded to students attending accredited academic or trade/technical institutions. To obtain an application, send a self-addressed, 9 X 12 envelope with $1.47 postage to ASFA/AMF Scholarship Program, 5211 Auth Rd., Suitland, MD 20746. Application forms are available from November 1 to march 31 annually.

Eligibility: The applicant must be an unmarried child or legally adopted stepchild of an enlisted member serving in the U.S. Air Force, Air national Guard or air Force Reserve Command, or in retired status, who has not attained their 23rd birthday as of August 31 of the award year and is dependent on the parent or guardian fro over one-half of his/her support.

Award(s) & Amount(s): Varies

❒ AMERICAN METEOROLOGY SOCIETY
AMVETS
4647 Forbes Blvd.
Lanham, MD 20706
Tel. (301) 459-6255

Description: AMVETS Auxiliary of the American Veterans of World War II, Korea & Vietnam Sponsors annual scholarship program for members and dependents.

Award(s) & Amount(S): Varies

❒ ASSOCIATION OF GRADUATES OF THE U.S. AIR FORCE ACADEMY
USAF Academy
3116 Academy Dr.
Colorado Springs, CO 80840-4475
Tel. (719) 472-0300

Description: Various scholarships offered to undergraduate/graduate children of the Air Force Academy graduates.

Award(s) & Amount(s): Varies

❒ BUDWEISER USO SCHOLARSHIP
USO World Headquarters
Scholarship Program
Washington Navy Yard
901 M. St., S.E.
Bldg. 198
Washington, DC 20374

Tel. (202) 610-5700

Deadline: March 1

Description: Undergraduate awards offered annually to dependents of active-duty service personnel of the U.S. Armed Forces.

Award(s) & Amount(s): 15, $1,000

❒ CAPTAIN JODI CALLAHAN MEMORIAL SCHOLARSHIP
Aerospace Education Foundation
1501 Lee Highway
Arlington, Virginia 22209-1198.
Tel. (703) 247-5800 ext. 4880.
E-mail: AEFStaff@aef.org

Deadline: July

Description: AEF's Captain Jodi Callahan Memorial Scholarship is awarded each year to an Air Force active duty member who is a member of the Air Force Association. This scholarship is for students pursing a Masters degree in a nontechnical field. The selection is based on the student's academic records, recommendations from an Air Force supervisor and an AFA official and demonstrated involvement with AFA. Applications are available at Base Education Offices, through AFA Chapter and State Presidents and Aerospace Education Vice Presidents and the AFA Fax Reply service (800) 232-3563 document # 860

Eligibility: Air Force active duty (enlisted or officer) Air Force Association members in good standing for 12 months prior to August 1, 1998. The member must be pursing a nontechnical Masters degree during off-duty time. Applicants must be enrolled in the current or upcoming semester with a minimum of 2 credit hours.

Note: Air Force rank: E-4, E-5, E-6 or E-7, CCAF Air Force graduate. May only receive the grant one time. Be enrolled or will be enrolled in an accredited program leading to a baccalaureate degree. Must submit application, narrative and documentation (see application) to a base Education Office.

Award(s) & Amount(s): $1,000

❒ DAEDALIAN FOUNDATION
P.O. Box 249
Randolph AFB, TX 78148-0249
Tel. (210) 945-2111

Deadline: Contact Sponsor

Provides scholarships and awards to ROTC, CAP and others who plan to become commissioned military aviators.

Award(s) & Amount(s): Contact Sponsor

❒ DISABLED AMERICAN VETERANS AUXILIARY
National Education Loan Fund Director
National Headquarters
3725 Alexandria Pike
Cold Spring, KY 41076
Tel. (606) 441-7300

Deadline: April 15

Description: No-interest loan funds available to Full Paid Life Members of the DAVA, or their dependents. Must be a full-time student in college, university or technical school.

Award(s) & Amount(s): Contact Sponsor

❒ NATIONAL WOMEN'S RELIEF CORPS (Auxiliary to the Grand Army of the Republic)
National Women's Relief Corps
629 South Seventh
Springfield, IL 62703
Tel. (217) 522-4373

Deadline: Contact Sponsor

Description: Applicants should be an undergraduate, descendant/relative of a member or a member of the Woman's Relief Corps. Submit requests for applications to the National Scholarship Committee through a local CORPS.

Award(s) & Amount(s): Varies

❒ NON-COMMISSIONED OFFICERS ASSOCIATION
P.O. Box 33610
San Antonio, TX 78265
Tel. (210) 653-6161

Deadline: Contact Sponsor

Description: Awards for undergraduate study or vocational education to dependents and spouses of current members.

Award(s) & Amount(s): Offers 35 awards annually of $750 to $1,000

❒ RESERVE OFFICERS ASSOCIATION
1 Constitution Ave, NE
Washington, DC 20002
Tel. (202) 479-2200

Deadline: Contact Sponsor

Description: College scholarships awarded to ROA members, children or grandchildren of ROA members or ROA Ladies Clubs.

Award(s) & Amount(s): 100 awards, up to $500

❒ THE RETIRED ENLISTED ASSOCIATION
1111 S. Abilene Ct.
Aurora, CO 80012
Tel. (303) 752-0660

Deadline: Contact Sponsor

Description: Dependents and grandchildren of members are eligible.

Award(s) & Amount(s): Offers 30, $1,000

❐ **RETIRED OFFICERS ASSOCIATION**
201 North Washington St.
Alexandria, VA 22314
Tel. (703) 549-2311

Description: Provides interest-free loans to undergraduate students who are dependent children of active, reserve and retired uniformed service personnel.

Award(s) & Amount(s): Varies

❐ **SOCIETY OF THE STRATEGIC AIR COMMAND**
Headquarters
P.O. Box 1244
Bellevue, NE 68005-1244

Contact: Ron Wink, (AF Ret. Col.), Education Chairman
Tel. (800) 952-2053 or (402) 293-7433
Fax (402) 292-5536

Deadline: Contact Sponsor

Eligibility: Full-time college sophomores, juniors or seniors who maintain a 3.0 GPA and participate in extracurricular activities or projects associated with their academic majors. Must be a U.S. citizen and sponsored by an active member of the society of the Strategic Air Command.

Award(s) & Amount(s): 2, $1000

❐ **VETERAN'S ADMINISTRATION**
Benefits Information Department
1120 Vermont Ave., N.W.
Washington, DC 20421
Tel. (800) 827-1000 or (202) 418-4343

Decryption: Information and referrals on a number of scholarship/loan/grant programs available to veterans.

Award(s) & Amount(s): Varies

SCHOOL SPONSORED

This section provides you with a listing of colleges, universities, and flight schools who sponsors their own awards. Students are awarded these scholarships, if they attend a specific school and meet the eligibility requirements. If you are interested in attending a specific school, look up the school in the school-specific section and there you will find a list of scholarships offered to their prospective and/or current students. This list is not inclusive, if you are interested in a particular school, contact the admissions office or aviation/aerospace department directly to obtain more information.

❒ COLLEGE OF AERONAUTICS GRANTS AND SCHOLARSHIPS

College of Aeronautics
LaGuardia Airport, Flushing, NY 11371
Tel. (800) PRO-AERO

For New Freshmen:

❒ Community Scholarship Program
A scholarship of $1,000 is awarded to entering freshmen who have been recommended by their guidance counselor. Scholarship applications are distributed to high school guidance counselors.

❒ Founders' Scholarship Program
Eligibility: A scholarship of $1,000 per year is awarded to an entering freshman who has been selected by the Director of Admissions. The applicant must have applied for this scholarship by submitting an application with an autobiographical essay.

❒ Freshman Academic Scholarships
Eligibility: High school seniors who have finished in the top 10, 20, and 30 percent in their class with SAT scores of above 1250, 1150, and 1050 respectively will qualify to receive scholarships in the amounts of $2,500, $1,750, and $1,750. The scholarships are renewable as long as the student meets the stated criteria.

❒ Leon D. Star, M.D. Memorial Scholarship
Eligibility: A resident of the Borough of Queens is chosen annually to receive a $2,500 scholarship by the JFK Chamber of Commerce in memory of the late director of the JFK Medical Center.

❒ Kiwanis Scholarship
Eligibility: The Kiwanis Club of LaGuardia Airport has established an annual scholarship in the amount of $1,500 for a graduate of Aviation High School to be applied towards tuition at the College of Aeronautics.

❒ Vocational Industrial Clubs of America (VICA) Scholarship
Eligibility: First, second and third place winners of the secondary VICA competition who enroll at the College of Aeronautics receives $3,000, $2,000 and $1,000 respectively in tuition scholarships.

New Transfer Students

❒ Transfer Scholarships
Graduates of two-year colleges, or students who have accumulated at least 60 credits at a four-year institution of higher education with grade point averages of 3.5 or higher are awarded scholarships in the amounts of $1,250 per academic year. Students with grade point averages of 3.0 or higher will receive $750 per academic year.

Continuing Students

❒ Academic Scholarship
Description: Scholarships are awarded annually to continuing students with outstanding academic records. Recipients are recommended by the College of Aeronautics Scholarship Committee and approved by the Vice President Chief Academic Officer.

❒ William Smart Alumni Association Scholarship
Description: This annual scholarship is given in the honor of William Smart, an early alumnus and faculty member.

❒ Board of Trustees Grants
Eligibility: Any student that demonstrates financial need may be considered for this grant in aid. An award of up to $850 per year is awarded based on need and may be awarded to students who have completed the financial aid process.

New York State Financial Aid Programs

❒ **Tuition Assistance Program (TAP)**
Description: This grant is available to full-time students who are New York state residents. Awards are based on New York state net taxable income and are used to pay for tuition.

❒ **Aid for Part Time Study (APTS)**
Eligibility: This grant is available to students who are enrolled for between three and 11 credits. These awards are based on New York state net taxable income and are available to New York residents to help offset tuition costs.

❒ **DELAWARE STATE UNIVERSITY**
Contact: Dr. Thomas McFlight, Chair
Airway Science Department
1200 North Dupont Highway
Dover, DE 19901-2277
Tel. (302) 739-3535
Fax (302) 739-7691
Website: www.DSC.EDU

Deadline: After Admission to the program

Descriptions: Seven flight scholarships are awarded each year to women training as pilots. It is given to those with serious purpose.

Eligibility: Applicant must be a woman enrolled in the Aircraft Systems major at Delta State University; 3.0 or better GPA

Award(s) & Amount(s): 7, $2200

Internships & Cooperative Education Programs: American Airlines, United Airlines, Northwest Airlines, FAA, Customs, Philadelphia International Airport, US Forest Service

---

❒ **EASTERN KENTUCKY UNIVERSITY**
Contact: Dr. Wilma J. Walker, Coordinator
Aviation Program, Burrier 404
Richmond, KY 40475-3107
Tel. (606) 622-1014
Fax (606) 622-1163
Website: http://www.eku.edu
E-mail: GEOWALKE@ACS.EKU.EDU

Deadline: Contact Sponsor

Eligibility: The aviation program offers five scholarships. All scholarships are awarded to students who have been enrolled at EKU for at least one semester.

Award(s) & Amount(s): 5, $500 - $1,500

❒ **FLORIDA INSTITUTE OF TECHNOLOGY**
Florida Tech Academic Scholarships
150 W. University Blvd.
Melbourne, FL 32901 - 6975
Tel. (407) 674-8030
Fax: (407) 723-9468
Website: http:www.fit.edu
E-mail: admissions@fit.edu

Contact: Michael Perry or Judi Marino, Office of Undergraduate Admissions
Tel. (800) 888-4348

Deadline: An application for admission is the only paper work needed. It must be received by February 1st, prior to the academic year in which the student wants to start. No scholarships are offered after April 1st.

Description: Florida Tech offers various tuition scholarships to incoming Freshman and transfer students. The scholarships are for full-time degree seeking students (includes flight training in degree). The scholarship amounts vary and are determined by the admissions office based on prior academic performance (GPA and SAT or ACT test scores). These scholarships are for Florida Tech Students only, and only one scholarship per student.

Eligibility: Students must apply and be accepted into a degree program at Florida Tech as a full-time student. For scholarship consideration, students must have a strong academic background, GPA and SAT 1 or ACT test results. Based level scholarships start with a GPA of 3.0/4.0 plus and SAT 1 - 1000 plus.

Award(s) & Amount(s): Vary, Range from $10,000 - $40,000 over a for year period ($2,500 - $10,000 per academic year).

Internships & Cooperative Education Opportunities: Yes

❒ FORT HAYS STATE UNIVERSITY
Financial Aid - Government Loans
Contact: Craig Karlin
Student Financial Aid
600 Park St.
Hays, KS 67601
Tel. (785) 628-4501
Fax (785) 628-4501
Website: http://FHSU.edu
E-mail: PHMU@FHSU.EDU

Deadline: Contact Sponsor

Eligibility: Full-time university students seeking a degree. Financial aid is available to students based on need and, for ground school and flight training courses as a part of another university degree program.

Award(s) & Amount(s): Varies

❒ GUILFORD TECHNICAL COMMUNITY COLLEGE
GTU Foundation
Contact: Casey Wallen
PO Box 309
Jamestown, NC 27282
Tel. (336) 334-4822
Fax (336) 665-9401
E-mail: fryee@Technet.gtcc.cc.nc.us

Deadline: December 1

Eligibility: Multiple scholarships, grants and loans are available to full-time students enrolled in flight training, A&P or management.

Award(s) & Amount(s): Varies

Internship & Cooperative Education Opportunities: Continental Airlines, Greensboro Airport Authority, Eastwinds Airlines

---

❒ INDIANA STATE UNIVERSITY AVIATION SCHOLARSHIPS
Aviation Association of Indiana
Dr. Roy Buckingham
Indiana State University
Aerospace Technology CL 116
Tel.: (812) 237-2660

The Department of Aerospace Technology at Indiana State University makes available the following scholarships to their students. Any student meeting the minimum qualification requirements may apply for one or more of these scholarships.

❒ **Aviation Association of Indiana Scholarship** Deadline: Early April
Eligibility: Be a graduate of an Indiana High School. Be enrolled in either: ISU Professional Pilot Program, ISU Aerospace Administration Program, Have junior standing in the spring semester, Have an average or better university academic record, Have participated in extracurricular activity or work activity, Not be a child of a member of the Aviation Association of Indiana, and Have an unmet financial need.

Award(s) & Amount(s): Flight major $1,000; Maintenance/Administration major $1,000

❒ **Capt. Ralph C. Miller Memorial Scholarship**
Eligibility: ISU Professional Pilot major; Enthusiastic attitude towards an aviation career; Accumulative grade point average of 2.5 or better upon application; Monetary award will be applied to the academic fees at Indiana State University for the following Academic year.

Award(s) & Amount(s): $300

❒ **Dennis J. Hunter Memorial Scholarship**
Eligibility: ISU Professional Pilot major; Accumulative grade point average of 2.5 or better upon application; Enthusiastic attitude towards an aviation career; Possess an FAA Private Pilot Certificate upon application and working on an advanced pilot certificate or rating. Preference given to those pursuing a multi-engine rating or CFI certificate; Monetary award will be applied to the academic fees at Indiana State University for the following Academic year

Award(s) & Amount(s): $500

❒ **Jeffrey Hardaway Memorial Scholarship**
Eligibility: ISU Aerospace Technology major; Enthusiastic attitude towards an aviation career; Completed between 48 and 112 semester hours by the end of the Spring semester in which application is made; Must have over 62 semester hours by the start of the following Fall semester; Accumulative grade point average of 2.7 or better upon application; Major grade point average of 3.0 or better upon application; Full time Aerospace student (12 semester hours or more) before and after issuance of the award; Financial need positively considered; Monetary award will be applied to the academic fees at Indiana State University for the following Academic year.

Award(s) & Amount(s): 2, $600 each

❒ **John A. Merritt Memorial Scholarship**
Eligibility: Completed sophomore year as an ISU Professional Pilot major; Accumulative GPA of 2.75 or better upon application; Major GPA of 3.0 or better upon application; Full-time student during issuance of scholarship; Monetary award will be applied to the academic fees at Indiana State University for the following Spring semester; Must perform between 10-15 hours of community service during the Fall semester, after selection for the award. A monthly progress report of service performed must be submitted to the Aerospace Department Chairperson. Monetary award will not be presented until the Spring semester and only after the 10-15 hours of community service have been performed and three progress reports have been accepted by the Aerospace Department Chairperson.

Award(s) & Amount(s): $350

❒ **Kenneth S. Papkoff Memorial Scholarship**
Eligibility: ISU Professional Pilot or two-year General Aviation Flight major; Accumulative grade point average of 2.5 or better upon application; Designed to assist worthy and deserving students; Renewable scholarship enabling recipient to reapply the following academic year; Monetary award will be applied to the academic fees at Indiana State University for the following Academic year

Award(s) & Amount(s): $750

❒ **Quentin R. Beecher Memorial Scholarship**
Eligibility: ISU Aerospace Technology major; Enthusiastic attitude towards an aviation career; Freshman or sophomore upon application; Full time Aerospace student (12 semester hours or more) upon; application; Accumulative GPA of 2.7 or better with a major GPA of 3.0 or better upon application; Possess an FAA Private Pilot Certificate upon application; Must pursue an FAA Commercial Pilot Certificate or higher; Monetary award will be applied to the academic fees at Indiana State University for the following Academic year.

Awards(s) & Amount(s): $250

❒ **IOWA CENTRAL COMMUNITY COLLEGE**
Aviation Program Scholarships
Contact: Elizabeth Fuller, Kelly Wirtz, or Ralph Storm
1725 Beach
Webster City, IA 50595
Tel. (515) 832-1632
Fax (515) 832-6315

Deadline: Contact Sponsor

Eligibility: This scholarships is available to students in flight training. Full-time students enrolled in courses to receive college credit.

Award(s) & Amount(s): $200 annual; $100 per semester

❒ **JACKSONVILLE UNIVERSITY**
Contact: Admissions Office
2800 University Blvd. North
Jacksonville, FL 32211
Tel. (904) 745-7000
Fax (904) 745-7021
Website: http://www.ju.edu
E-mail: admissions@ju.edu

Deadline: February 1 - same year of enrollment

Eligibility: Multiple scholarships are available to accepted Jacksonville University students. Some scholarships are based on financial need, academics, merit, and community service. Higher ranging and full tuition scholarships are available to students with a GPA of 3.25 with a SAT score of 1100 or greater.

Award(s) & Amount(s): $500 - $15,000 per year

Internship & Cooperative Education Opportunities: American Airlines, Comair Airlines, Comair Aviation Academy, Comair Jet Express, Northwest Airlines, US Airways, JAXPort, Denver International Airport, Paine/Boeing Field, NASA Aviation /Space Camp

---

❒ **LEWIS UNIVERSITY**
Contact: Ms. Clarie Teramerman
Rt. 53
Romeoville, IL 60446
Tel. (800) 897-9000

Deadline: Contact Sponsor

Description: There are three scholarships offered to students enrolled in any aviation major at Lewis University. Scholarship are as follows: the Harold E. White Endowed Scholarship, Scott Jasinski Memorial Scholarship, Charie Walls Maintenance Memorial Scholarship.

Eligibility: Must be accepted/enrolled at Lewis University; Residents of Naperville, IL or DuPage County, or the State of Illinois.

Award(s) & Amount(s): Varies

Internships & Cooperative Education Opportunities: Yes

❐ MIAMI - DADE COMMUNITY COLLEGE
Eig - Watson Scholarship (Sponsored by Lois & Harvey Watson)
Contact: Yoel Hermandez or Marilyn Kern - Ladner
Homestead Campus
500 College Terrace
Homestead, FL 33030
Tel. (305) 237 - 5060
Fax: (305) 237 - 5135
Website: http://www.ipcc.com/avweb-prep/miami_dade
E-mail: aviation@mdcc.edu

Deadline: March

Eligibility: Awarded to students pursuing a Professional Pilot Technology Degree. U.S. Citizen, passes a Private Pilot Certificate; Must be a Miami - Dade Community College student enrolled in the Professional Pilot Program.

Award(s) & Amount(s): 10, Ranges $3,500 to $5,000

❐ KANSAS STATE UNIVERSITY
"Dolly" Hardman Scholarship
223 College Center
Salina, KS 67401
Tel. (785) 826-2638
Fax: (785) 826-2936
Website: http://www.ksu.edu

Deadline: February

Description: Marion W. "Dolly" Hardman memorial scholarship is used for flight training costs and other educational needs. The amount may vary depending on the number of applicants.

Eligibility: Female, planned enrollment in the flight program at Kansas State University - Salina. Must meet two of the following criteria - composite ACT score of 25, a 3.5 GPA in high school, top 15% of the graduating class or be the salutatorian or valedictorian of their graduating class.

Award(s) & Amount(s): Varies

❐ MIDDLE TENNESSEE STATE UNIVERSITY
MTSU - Aerospace Department Scholarships

Listing and qualifications for applying and receiving a scholarship in the aerospace department are as follows: Normally all applicants must be enrolled in the aerospace program here at MTSU. However, there are a few available to incoming freshmen, that are required to major in aerospace. Applications must be sent in before June 1and award will be before 30 June each year.

❐ Aerospace Graduate Scholarship
Eligibility: The scholarship amount to be determined by department chair. Applicants must include need and have an excellent GPA (3+) and either be on internship or other work related activities and formally accepted for graduate work by the aerospace department.

Amount: Contact Sponsor

❐ **Airshow Aviation Scholarship**
Description: This scholarship is sponsored by The Smyrna, Lavergne and Donelson Rotary Club. Applicants are to be aerospace majors and residents of Tennessee and be a sophomore, junior or senior. Applications are to be typed in letter form, stating how this scholarship will be used and a copy of latest transcript.

Amount(s): $500 each semester

❐ **Colonel Jean Jack Aerospace Scholarship**
Description: Awarded in the fall or spring semesters established in honor of Colonel Jack's contribution to the aerospace department and aviation in general. Applicants are to be aerospace majors and be a sophomore, junior or senior. Applications must be typed in letter form outlining how this scholarship will be used and a copy of latest transcript.

Amount: $500

❐ **Col. Harry E. Slater Memorial Scholarship**
Eligibility: Established by the 94th Bomb Group Association, in honor of Colonel Slater. Available Fall semester each year to a student majoring in aerospace and is a junior or senior with an excellent grade point average. Application for scholarship must be typed and indications of how it will help you plus a copy of latest transcript.

Amount(s): $900

❐ **Dr. Wallace R. Maples Aerospace Scholarship**
Eligibility: Established in honor of the aerospace department former chairman. Available spring semester each year to student majoring in aerospace and may be a sophomore, junior or senior with an excellent grade point average. Application for scholarship must be typed and indications of how it will help you plus a copy of latest transcript.

Amount: $500

❐ **Don Ace Memorial Aerospace Scholarship**
Eligibility: Established in honor of his contribution to aviation in Middle Tennessee. Available spring semester each year to students majoring in aerospace (professional pilots) and is a sophomore, junior or senior, with an excellent grade point average. Application for scholarship must be typed and indicate how this will help you in your pilot training, plus a copy of your latest transcript.

Amount: $750

❐ **Frank & Harriett Hedrick Memorial Aviation Scholarship**
Description: Established in honor of frank and Harriett Hedrick, a vice president of Beech Aviation. Available every spring semester to sophomore, junior and senior majoring in aerospace. Application must be typed in letter form indicating how it will be used plus copy of latest transcript.

Amount: $500

❐ **Freshmen Scholarship to the Aerospace Program**
Eligibility: Student must have an outstanding high school academic record, SAT or ACT. The amount of scholarship to be determined at the time of award. The department expects to give 3 scholarships annually. Applications must be typed in letter form indicating the need and desire for an aviation education.

❒ **H. Miller Lanier Memorial Aviation Scholarship**
Description: This scholarship has been established in honor of H. Miller Lanier's contributions to the state aviation activities. Applicants are to be aerospace majors and be a sophomore, junior or senior. Applications are to be typed in letter form, stating how this scholarship will be used and a copy of latest transcript. Award(s) &

Amount(s): $500 for the fall or spring semester,

❒ **Jeffrey Clayton McCrudden Memorial Aviation Scholarship**
Description: Mr. McCrudden was a student in the MTSU Aerospace Department. Available fall semester. Must be a professional pilot, completed private pilot requirements and enrolled in Aerospace 30GB Instrument Flight Instruction II and related activities leading 5 his/her instrument rating and commercial pilot certificate, GPA 2.5 minimum and recommended by their flight instructor. Typewritten application stating career goals, qualifications and aspirations and a copy of transcript.

Amount: minimum $500 (estimated)

❒ **Jim Price Jr. Memorial Scholarship**
Description: This scholarship is established in honor of Jim Price a Federal Express pilot and general aviation enthusiast. Applicants must be aerospace majors, minimum 3.2 GPA, with emphasis as a professional pilot. Must write a paper on "why I have chosen a career in aviation". Should be a junior or senior with one year remaining and copy of transcript. All applications must be typewritten. Award will be for fall/spring academic year.

Amount(s): $1000—$2000 each semester

❒ **Metropolitan Nashville Airport Authority Aviation Scholarship**
Eligibility: Applicant must be an aerospace major with an excellent grade point average. Should be interested in working at airports or managing them. Application must be typed and stating career goals and aspirations and a copy of latest transcript. Amount of scholarship to be determined at beginning of academic year.

Amount: Contact Sponsor

❒ **Southeastern Airport Managers Assos/American Association Airport Executives Southeast**
Description: MTSU has been selected to receive this scholarship for the past 15 years as one of the colleges and universities supporting aviation activities in the South. Available each school year. Applicants must obtain application forms and must be typed and also meet a scholarship board. Applicant must be junior or senior with one year remaining.

Amount: $750 — $1500/each semester

❒ **NORTHWESTERN MICHIGAN COLLEGE**
Contact: Judy Monaco
1701 E. Front St.
Traverse City, MI 49686
Tel. (616) 922-1220
Fax (616) 929-7116
Website: nmc.edu
E-mail: jmonaco@nmc.edu

❒ **Aviation Division Scholarship** Deadline: March 20
Eligibility: Completed 20 credits at NMC (which may include Spring semester); Completed a minimum of 5 aviation credits at NMC; 3.0 minimum GPA at NMC (3.25 minimum GPA in NMC aviation courses); Enroll as a full-time aviation student for the academic year and take a minimum of 12 credits each semester (Fall & Spring) including 6 aviation credits during the academic year; Demonstrate outstanding scholarship, airmanship, and service to the Aviation Division; Oriented toward an aviation career; Preference will be given to students have a

Private Pilot Certificate and who started their aviation training at NMC; Presidential Scholars and NMC Flight Instructors are not eligible; Scholarship is not based on financial need

Submit the following documents to Mike Stock, Chair of the Aviation Division Scholarship Committee: Scholarship Application Form (copies at Dispatch), Personal Essay (as described on Scholarship Application), and Personal Resume. A personal interview before the Scholarship Committee may be required (You will be contacted if an interview is required)

Award(s) & Amount(s): $2000 for use during the academic year, divided equally between semesters

❒ **The Frank P. Macartney Foundation** Deadline: July 1
Description: This scholarship is awarded to outstanding students with a passion for aviation excellence. Scholarships will be awarded to future aviators with high academic and technical achievement attending Northwestern Michigan College

Eligibility: Applicant must be a U.S. Citizen. Applicant must be a sophomore, junior or senior currently enrolled in a two-year or four-year aviation program with plans to complete a curriculum leading to a degree and a career in the field of aviation. Applicant must have achieved a minimum grade point average of *3.25* (on a 4.0 system) at the time of application. Previous recipients are not eligible.

Award(s) & Amount(s): $1000

❒ **Holts Claw Memorial Scholarship** Deadline: September 30

Eligibility: The applicant must be a second-year aviation student with at least a 3.0 grade point average. (Hold the Private Pilot License and have completed 24 semester credit hours.). The applicant must submit a resume and a 300 word essay explaining your background, goals in aviation, and financial need. Submit the resume and essay to Bob Buttleman. The applicant will be a full time student working towards his/her ratings. (Currently taking 12 semester credits or more.) A demonstrated background of an interest in flying plus leadership capabilities. A career goal of flying with an airline.

The scholarship is to be used to pay for Pilot Training flight fees at Northwestern Michigan College only. If the student transfers or leaves NMC, the unused funds will be returned to the Holtsclaw Scholarship Fund.

Award(s) & Amount(s): $2500

❒ **OKLAHOMA STATE UNIVERSITY**
Mary Francis Blair Memorial Scholarship
Contact: Glen Nemeciek
318 Willard Hall
Stillwater, OK 74078
Tel. (405) 744-8062
Fax (405) 744-7758

Deadline: March 31

Eligibility: Full-time students with a minimum of 2.5 GPA attending Oklahoma State University

Award(s) & Amount(s): $1,000

❒ **SOUTHERN ILLINOIS UNIVERSITY AT CARBONDALE**
Aviation Management. & Flight
College of Applied Sciences & Arts
Carbondale, IL 62901
Tel. (618) 453-1147 ext. 238

❒ **Cessna, Gatewood, and Staggerwing** Deadline: April
Description: Three scholarships (Cessna, Gatewood, and Staggerwing) are awarded to assist students who are currently enrolled in the Aviation Flight Associate of Applied Science degree program, through the College of Applied Science and Artd at Southern Illinois University at Carbondale. The money for the Staggerwing scholarship will be awarded in the Spring semester and the Gatewood and Cessna scholarships will be awarded in the Fall semester.

Eligibility: Primary consideration will be based on the student's academic standing, financial need, and advancement in the flight training program. Students must have one semester of flight classes taken at Southern Illinois University before applying.

Award(s) & Amount(s): Contact Sponsor

❒ **Jerry Kennedy Aviation Career Advancement Scholarship**
Eligibility: For current SIUC Aviation Management and Aviation Flight students only! The applicant must have completed two years toward degree at SIUC with minimum overall GPA of 3.0 in course work; be currently involved in advanced Flight Training (Instrument, Commercial, Flight Instructor, or Multi-Engine); demonstrate a need for financial assistance. Deadline: October 17

Award(s) & Amount(s): Varies

❒ **William R. Norwood Aviation Scholarship** Deadline: October
Description: The William R. Norwood Scholarship is established to encourage other to excel in aviation. it is bestowed upon those who demonstrate a strong commitment, devotion, and sacrifice to aviation in the William R. Norwood tradition.

Eligibility: Evidence of participation in aviation related organizations and activities; Minimum of a 2.75 overall GPA; Must be a sophomore, junior, or senior enrolled at Southern Illinois University at Carbondale in the aviation flight program.

Award(s) & Amount(s): 1, $1,000

❒ **PARKS COLLEGE OF ST. LOUIS UNIVERSITY COBRO MAINTENANCE SCIENCE SCHOLARSHIPS**
COBRO Maintenance Science Scholarship
Parks College of St. Louis University
Cahokia, IL 62203

Award(s) & Amount(s): Up to $10,000, or 120% of a student's annual tuition and fees, up to a 3 year minimum of $30,000.

❒ **TEXAS SOUTHERN UNIVERSITY**
Wings Over Houston Airshow Scholarship
Wings Over Houston Airshow Festival
Contact: Prof. J. Richmond Nettey or Col. Jack Amuny
6284 Brookhill Dr.
Houston, TX 777087
Tel. (713) 644-1018
Fax: (713) 644-3479
Website: http://www.avdigest.com/woh
E-mail: tchtirnettey@tsu.edu

Deadline: Contact Sponsor

Description: The scholarships are intended to assist and reward students who have demonstrated academic potential, leadership, and extra curricular involvement. Student must attend Texas Southern University.

Eligibility: At least 60 semester credit hours completed in an aviation or aviation related major, 3.0 GPA, full-time student, U.S. citizen/permanent resident, legal resident in Harris County or surrounding counties (Galveston, Brazoria, Fort Bend, Waller, Montgomery, Liberty, and Chambers).

Award(s) & Amount(s): At least $1,000

❒ UNIVERSITY OF CINCINNATI, CLERMONT COLLEGE
Joseph Vorbeck Scholarship
Contact: Shirley Quinn or Ben Roller at (513) 732 - 5284
4200 Clermont College Dr.
Batavia, OH 45103
Tel. (513) 735 - 9100 ext. 298
Fax: (513) 735 - 9200
Website: http://www.sportys-catalogs

Deadline: March

Eligibility: Two scholarships are awarded to incoming Freshmen and two will be awarded to Sophomores. Students must be enrolled in the Professional Pilot Major at the University of Cincinnati, Clermont College.

Award(s) & Amount(s): 4 scholarships, $2,500

❒ UNIVERSITY OF ILLINOIS - INSTITUTE OF AVIATION
J.W. Stonecipher Scholarship
Contact: Dr. T.W. Emanuer
1 Airport Rd.
Savoy, IL 61874
Tel. (217) 244-8671
Fax (217) 244-8761
Website: http://www.aviation.uiuc.edu
E-mail: emanuel@uiuc.edu

Deadline: Contact Sponsor

Eligibility: Three scholarships are awarded to students enrolled in the Professional Pilot Program at the University of Illinois.

Award(s) & Amount(s): 3, $1,000

Internship & Cooperative Education Opportunities: United Airlines, American Airlines, Trans World Airlines

---

❒ UNIVERSITY OF MARYLAND EASTERN SHORE
Contact: Dr. Marc B. Wilson, CPE
Aviation Science Scholarship
Engineering & Aviation Department
30806 University Blvd., South
Princess Anne, MD 21853
Tel. (410) 651 - 6365
Fax: (410) 651 - 7946 or 7959
Website: http://www.umes.edu
E-mail: mwilson@umes-bird.umd.edu

Deadline: Contact Sponsor

Eligibility: Ant aviation science concentration. Must have a 3.0 or higher the preceding semester. A full-time student in the Aviation Science at University of Maryland Eastern Shore.

Award(s) & Amount(s): Varies

Internships & Cooperative Education Opportunities: FAA, NASA, Piedmont Airlines, Salisbury Airport

❒ UNIVERSITY OF MINNESOTA

Contact: Donna Rosenthal
Administrative Director
Aerospace Engineering and Mechanics
107 Akerman Hall
110 Union Street S.E.
University of Minnesota
Minneapolis, MN 55455
Tel. (612) 625-8000

There are several awards offered to students attending the University of Minnesota. All of these awards are for University of Minnesota students only. Provided below is a brief description of each award. Contact the university for more information.

❒ Boeing Scholarship
Application Deadline: Announced each year along with application procedures.
Description: An undergraduate scholarship for a student majoring in aerospace engineering at the University of Minnesota and who is a citizen of the U.S. It is awarded to a full-time student entering their junior year who is classified as an upper-division student. Receipt of the award is based on scholastic merit and high potential. The recipient must be authorized to work on a full-time basis in the United States and have an expressed interest in employment with the Boeing Company. Renewals of scholarships will be evaluated and determined annually. This award should not be assumed to be renewable for an individual's senior year.

Application Process: Will be announced prior to visit to campus by Boeing Representatives. Generally, students are required to complete an application form and submit a short resume or attend a career fair hosted by the Boeing Representatives at the request of the AEM Department. The Boeing Representatives make the final decision regarding scholarship recipient(s). NOTE: The recipient(s) may be required to complete an internship either prior to or after receiving the scholarship. (depending upon availability of funding)

Award(s) & Amount(s): 1 to 2, $2,000 per academic year

❒ I.T. Undergraduate Research Assistantship Program
Application Deadline: Not applicable; selection taken from incoming freshman applicants to the Institute of Technology.

Description: These scholarships and fellowships are awarded to incoming freshman who have designated Aerospace Engineering and Mechanics as their projected major and who have indicated on the I.T. Honors application an interest in receiving an I.T. Undergraduate Research Assistantship . Decisions are based on the complete U of MN application and on the I.T. Honors application.

Award(s) & Amount(s): Two awards are given through Aerospace Engineering and Mechanics: The Chester Gaskell Aeronautical -- Engineering Scholarship, Minnesota Space Grant Consortium Fellowship

Both awards provide the recipients the opportunity to participate in
laboratory or other research in Aerospace Engineering and Mechanics under the guidance of a faculty sponsor. For the Minnesota Space Grant Consortium Fellowship, the recipient must be a U.S. Citizen.

Award(s) & Amount(s): 1 (both) $1,000 (both)

❒ R. Minkin Aerospace Engineering Scholarship
Description: An undergraduate scholarship for a student majoring in Aerospace Engineering and Mechanics at the University of Minnesota. It is awarded to a sophomore student (72 credits completed satisfactorily) with the best GPA after five consecutive quarters. Completion of an application is not required. The Aerospace Engineering and Mechanics Awards Committee reviews each eligible undergraduate's GPA and makes their selection annually.

Award(s) & Amount(s): 2 to 3, $500

❒ UNIVERSITY OF OKLAHOMA - AVIATION DEPARTMENT

Joe Coulter Aviation Scholarship
Contact: Glenn Schaumburg
1700 Lexington
Norman, OK 73069
Tel. (405) 325-7231
Fax (405) 325-0136

Deadline: Unknown

Eligibility: Two scholarships are awarded each semester to students enrolled in flight training at the University of Oklahoma - Aviation Department.

Award(s) & Amount(s): 2, $500

❒ WINONA STATE UNIVERSITY

Frankard Family Scholarship
Contact: George Bolon
PO Box 5838
Winona, MN 55987
Tel. (507) 407 - 5585

Eligibility: Available to students who have completed one semester and enrolled in the Airway Science Program. Must have a 3.0 GPA or higher, good academic and social standings - Airway Program participant is not limited to these factors.

Award(s) & Amount(s): Ranges $350 to $2,000

Internships & Cooperative Education Opportunities: Metropolitan Airport Commission, Minneapolis Department of Transportation (MN DOT), TIC/MCC

SPECIAL INTEREST & AFFILIATION

❒ AAAA FRESHMEN SCHOLARSHIPS

Army Aviation Association of America Scholarship Foundation
49 Richmondville Avenue
Westport, CT 06880-2000
Tel. (203) 226-8184
Fax: (203) 222-9863

Deadline: April

Purpose: To provide financial aid for the post-secondary education of members of the Army Aviation Association of America (AAAA) or of their dependents.

Eligibility: This program is open to AAAA members and their siblings or children. Applicants must be high school seniors or graduates. They must be enrolled or accepted for enrollment as freshmen at an accredited academic institution. Special consideration is given to applications submitted or sponsored by warrant officers or enlisted personnel. Selection is based on academic merit and personal achievement.

Award(s) & Amount(s): At least 40 each year; of those, 1 for $10,000 ($2,500 per year) is reserved for a student pursuing a 4-year degree in engineering; another, for $4,000 ($1,000 per year) is designated for an applicant pursuing a 4-year degree in an aeronautical-related science. Scholarships are at least $1,000. Some are as high as $3,000 per year.

❐ AAAA GRADUATE SCHOLARSHIPS
Army Aviation Association of America Scholarship Foundation
49 Richmondville Avenue
Westport, CT 06880-2000
Tel. (203) 226-8184
Fax: (203) 222-9863

Deadline: April

Purpose: To provide financial aid for the graduate education of members of the Army Aviation Association of America (AAAA) or of their dependents.

Eligibility: This program is open to AAAA members and their siblings or children. They must be graduate students in an accredited college or university. Special consideration is given to applications submitted or sponsored by warrant officers and enlisted personnel. Selection is based on academic merit and personal achievement.

Award(s) & Amount(s): 3 each year. Scholarships are $1,000 per year.

❐ AAAA SPOUSE SCHOLARSHIPS
Army Aviation Association of America Scholarship Foundation
49 Richmondville Avenue
Westport, CT 06880-2000
Tel. (203) 226-8184
Fax: (203) 222-9863

Deadline: April

Purpose: To provide financial aid for the post-secondary education of spouses of Army Aviation Association of America (AAAA) members.

Eligibility: This program is open to the spouses of members of the AAAA who are pursuing college studies on the undergraduate or graduate level. Selection is based on academic merit and personal achievement.

Award(s) & Amount(s): Varies; generally, at least 2 each year. At least $1,000 per year.

❐ AAAA UPPERCLASSMEN SCHOLARSHIPS
Army Aviation Association of America Scholarship Foundation
49 Richmondville Avenue
Westport, CT 06880-2000
Tel. (203) 226-8184
Fax: (203) 222-9863

Deadline: April

Purpose: To provide financial aid for the post-secondary education of members of the Army Aviation Association of America (AAAA) or of their dependents.

Eligibility: This program is open to AAAA members and their siblings or children. They must be at least sophomores in college. Special consideration is given to applications submitted or sponsored by warrant officers and enlisted personnel. Selection is based on academic merit and personal achievement.

Award(s) & Amount(s): 6 each year; Scholarships are $1,000 per year.

❒ AIR LINE PILOTS ASSOCIATION SCHOLARSHIP
Air Line Pilots Association (ALPA)
Scholarship Committee
1625 Massachusetts Avenue
Washington, DC 20036
Tel. (703) 689-2270 or (703) 639-4265

Deadline: April 1

Description: Sons or daughters of medically retired or deceased pilot members of the Air Line Pilots Association. Although the program envisions selection of a student enrolling as a college freshman, eligible individuals who are already enrolled in college may apply. The student must be enrolled in a program of studies principally creditable toward a baccalaureate degree.

Eligibility: ALPA offers a four-year scholarship to sons and daughters of deceased or medically retired ALPA pilots who demonstrate academic achievement and financial need. One 4-year undergraduate college scholarship is awarded each year. The total monetary value is disbursed annually to the recipient for four consecutive years, provided that the adequate academic standing is maintained (3.0 GPA).

Amount(s) Awarded: 1, $12,000 ($3,000 given annually for four years of college)

❒ ALPHA ETA RHO SCHOLARSHIP
Alpha Eta Rho Aviation Fraternity
Alpha Eta Rho National Headquarters
Dr. Kent Backart
1615 Gamble Lane
Escondido, CA 92029

Deadline: March 9

Scholarship Criteria: (1) Applicants must be active members of a local Alpha Eta Rho Chapter and recommended by the Chapter (only one applicant is accepted from each Chapter); and (2) Need a minimum GPA of 3.0. Notes: Include application form and letter indicating future plans.

Award(s) & Amounts(s): 6, $500

❒ AMERICAN GEOLOGICAL INSTITUTE MINORITY PARTICIPATION PROGRAM SCHOLARSHIPS
Tel. (703) 379-2480

Description: Open to African-Americans, Native Americans, and Hispanic Americans.

Eligibility: For undergraduate or graduate students who study in the following fields: Earth Sciences, Space Sciences, or Marine Sciences. Must be a U.S. citizen.

Award(s) & Amount(s): Approximately 50 awards per year. Renewals possible

---

❒ AVIATION SCHOLARSHIP
Organization of Black Airline Pilots, Inc.
P.O. Box 5793
Englewood, NJ 07631
Tel. (201) 568-8145

Purpose: To provide financial assistance to Black teenagers and other minorities who are interested in aviation.

Eligibility: Black teenagers and other minorities who are interested in learning how to fly are eligible to apply.

Special features: The Organization of Black Airline Pilots was established to make certain Blacks and other minorities had a group that would keep them informed about opportunities for advancement within commercial aviation. The organization co-sponsors this program with the Summer Flying School in Tuskegee, Alabama.

Award(s) & Amount(s) Varies

❑ CAPTAIN GRANT T. DONNELL & ADMIRAL JIMMY THACH MEMORIAL SCHOLARSHIPS
Tailhook Foundation
P.O. Box 26626
San Diego, CA 92196-0626
Tel. (800) 269-8267

Deadline: July

Purpose: To provide financial assistance for post-secondary education to dependents of Tailhook Association members.

Eligibility: This program is open to the dependents of Tailhook Association members and to the dependents of individuals who have served as Navy, Coast Guard, or Marine Corps personnel aboard aircraft carriers. Applicants may be high school seniors, high school graduates, college students, or graduate students. They should be interested in working on a degree in aerospace education. Selection is based on educational achievements, merit, citizenship, and financial need.

Limitations: The future status of this program is uncertain.

Award(s) & Amount(s): 6, $1,500 (stipend), Duration: 1 year

❑ ILLINOIS PILOTS ASSOCIATION SCHOLARSHIP
Illinois Pilots Association
Contact: Ruth Frantz
46 Apache Lane
Huntley, IL 60142
Tel. (847) 669-3821
Fax: (847) 669-3822

Deadline: April

Eligibility: Scholarship is awarded annually to a full-time student enrolled in an Illinois college/university aviation degree program (not flight training).

Note: Must be an Illinois resident.

Award(s) & Amount(s): 1, $500

❑ NASA UNDERGRADUATE STUDENT RESEARCHERS PROGRAM
Contact: Robert Lawrence
Tel. (216) 433-2921
Fax: (216) 433-2348

Description: The program is administered through NASA Training Grants to participating institutions. The total amount of award varies based on financial need. Students may receive up to a maximum $12,000 for tuition assistance, other academic costs, and a stipend for summer research.

Eligibility: The NASA Undergraduate Student Researchers Program attempts to increase the number of students from minority institutions pursuing undergraduate degrees in selected areas of science and engineering compatible with NASA's mission in aeronautics, space science and aerospace technology.

Award(s) & Amount(s): Varies

❒ NAVY TUITION ASSISTANCE PROGRAM

U.S. Navy
Attn.: Naval Education and Training Program Management Support Activity
6490 Saufley Field Road
Pensacola, FL 32509-5204
Tel. (904) 452-1806
Toll Free (800) USA-NAVY

Purpose: To provide financial assistance for the high school, vocational, undergraduate, or graduate education of Navy personnel.

Eligibility: Active duty Navy personnel (both officers and enlisted people, including Naval Reservists on continuous active duty or ordered to active duty for 120 days or more) who register to take courses at accredited civilian schools during off-duty time are eligible for this program. Tuition assistance is provided for courses taken at accredited colleges, universities, vocational/technical schools, private schools, and through independent study. Recipients must maintain at least a "C" average for undergraduate courses and "B" average for graduate courses.

Limitations: Officers must agree to remain on active duty for at least 2 years after completion of courses funded by this program. Tuition assistance may not be used for flight training, or for physical education and recreation courses (unless the sailor is pursuing a physical education/recreation major or the courses are required for degree completion).

Duration: Until completion of a Bachelor's or graduate degree.

Award(s) & Amount(s): Varies each year. Navy personnel chosen for participation in this program continue to receive their regular Navy pay. The Navy will also pay 100 percent of tuition for high school completion courses; 75 percent of a maximum of $125 per credit hour (to a maximum of $285 per course) for undergraduate courses; 75 percent of a maximum of $175 per credit hour (to a maximum of $395 per course) for graduate courses; and 75 percent of tuition (to a maximum of $1,300 per person per fiscal year) for vocational or technical courses.

❒ NHCFAE COLLEGE STUDENT SCHOLARSHIPS

National Hispanic Coalition of Federal Aviation Employees
2300 East Devon Avenue
Des Plaines, IL 60018
Tel. (312) 694-7893

Deadline: April

Purpose: To provide financial assistance to minority and women students who are interested in careers in aviation.

Eligibility: Eligible to apply for this support are minority and women students pursuing full-time postsecondary education leading to a career in aviation. Applications include a research paper on a given topic relating to aviation, a financial need statement, and a personal biography that covers academic standing and career objectives.

Special features: The National Hispanic Coalition of Federal Aviation Employees is a nonprofit organization comprised mainly of Hispanics and other minorities who are employed at the Federal Aviation Administration. Further information on these scholarships is available from Diana Lopes-Story, Director of Women's Initiative, P.O. Box 697, Newcastle, OK 73065.

Award(s) & Amount(s): 2 each year. The stipend is $500 per year; funds are paid directly to the college or university. Duration: 1 year; may be renewed.

❒ NEBRASKA SPACE GRANT SCHOLARSHIPS AND FELLOWSHIPS

Nebraska Space Grant Consortium
UNO Aviation Institute
Allwine Hall 422
University of Nebraska at Omaha
Omaha, NE 68182-0508
Tel. (402) 554-3772
Fax: (402) 554-3781
E-mail: nasa@cwis.unomaha.edu

Purpose: To fund aerospace-related research and studies on the undergraduate and graduate school level for students in Nebraska.

Eligibility: This program is open to all eligible undergraduate and graduate students at the following schools in Nebraska: University of Nebraska at Omaha, University of Nebraska at Lincoln, University of Nebraska at Kearney, University of Nebraska Medical Center, Creighton University, Western Nebraska Community College, Chadron State College, and Nebraska Indian Community College. Applicants must be U.S. citizens and working on a degree in an aerospace-related area. Special attention is given to applications submitted by women, underrepresented minorities, and individuals with disabilities.

Special features: Recipients conduct research in an aerospace-related area and receive at least 3 semester credits for that activity during the year of the award. Funding for this program is provided by the National Aeronautics and Space Administration.

Limitations: Recipients must submit a progress report each semester on the aerospace project to their designated faculty monitor. Failure to provide that report disqualifies the student from reapplying for a renewal fellowship.

Duration: Academic awards are 1 year; summer awards are for the summer months. Both awards are renewable.

Award(s) & Amount(s): Academic year awards are $7,500; summer awards are $2,500.

❒ OREGON EDUCATIONAL AID FOR VETERANS

Oregon Department of Veterans' Affairs
Attn.: Veterans' Services Division
700 Summer Street N.E., Suite 150
Salem, OR 97310-1270
Tel. (503) 373-2085
Toll Free (800) 692-9666
TDD: (503) 373-2217

Purpose: To provide financial assistance for the post-secondary education of certain Oregon veterans.

Eligibility: Veterans eligible for this aid are those who served on active duty in the U.S. armed forces for not less than 90 days during the Korean War or received an Armed Forces Expeditionary Medal or the Vietnam Services Medal for services after July 1, 1958, were released under honorable conditions, were Oregon residents for 1 year immediately before entering the service and when applying for aid, and are U.S. citizens.

Special features: Benefits are paid for classroom instruction needed as part of an apprenticeship program or other on-the-job training program. Benefits are also paid for home study courses and for vocational flight training.

Limitations: Educational Aid will not be paid if the veteran is receiving federal GI training benefits for that course. School officials are required to certify the amount that the student paid for tuition, lab fees, books, and supplies, and payments for each month or portion of a month will be made only if the actual cost of the course is equal to or greater than the payment amount.

Duration: Benefits are paid for as many months as the veteran spent in active service, up to a maximum of 36 months. One month of entitlement will be charged for each month paid, regardless of the amount paid.

Award(s) & Amount(s): Varies each year. Full-time undergraduate college students are entitled to receive up to $50 per month; part-time college, vocational, and graduate students are entitled to receive up to $35 per month; students enrolled in home study college courses receive a refund of the cost of the course, not to exceed $35 for each month the course is taken.

❒ PALWAUKEE AIRPORT PILOTS ASSOCIATION

1120 S. Milwaukee Ave., Suite A
Wheeling, IL 60090 - 6392
Tel/Fax: (847) 265- 9391

Deadline: June

Description: The Palwaukee Airport Pilots Association wish to award one scholarship to an **individual who resides in Illinois**, currently attends an accredited University, College or Aviation Technical School located in the state of Illinois, and is pursuing a course of study in a recognized professional aviation program.

Eligibility: Applicants must be full-time students with freshman, sophomore or junior year standing. Scholastic standing, indicative of the student's achievements, is a minimum GPA of 3.7 based on a 5.0 grade point system. Involvement in school and community activities as well as activities and experiences related to aviation shall demonstrate the applicant's motivation. The need for financial assistance shall be substantiated.

Award(s) & Amount(s): 1, $1,000

❒ RHODE ISLAND PILOTS ASSOCIATION

RIPA Scholarship Chairman
Marilyn Biagetti
104 East Ironstone Road
Harrisville, RI 02830

Deadline: February 20

Eligibility: Applicant must be at least 16 years of age**. Applicant must be a resident of Rhode Island.** Money awarded must be used towards goals in aviation. Applicant must be able to pass a third class physical. If under the age of 18 years old, applicant must have written permission of their parents/guardian. A personal letter (not over 1 1/2 pages in length) describing how the scholarship money will be used and aviation goals. A list of extra curricular activities, hobbies, etc. if any. A letter of recommendation from a person having knowledge of the applicant's goals and activities. All requirements must be met.

Award(s) & Amount(s): Contact Sponsor

❒ ROBBINS AIRPORT SCHOLARSHIP

Aviation Scholarship Foundation
P.O. Box 246
Palos Park, IL 60464
Tel. (708) 448-1914
Website: http://members.aol.com/kidsfly
E-mail: kidsfly@aol.com

Deadline: April 15

Eligibility: A fully funded private pilot program for eighth grade and freshmen students from Chicago's south suburbs. Low-income youths that reside in Robbins, Harvey, or Ford Heights are favored. The glider program is the entry level for any student. Students who successfully complete the glider program by earning their private pilot's certificate are invited to began the airplane program. Students fly every Saturday morning and work towards their private pilot certification in the gliders and then the single-engine airplane. All costs are covered through certification. Students usually fly gliders for two seasons and then the airplane for a year.

Award(s) & Amount(s): Contact Sponsor

❐ **SOCIETY OF FLIGHT TEST ENGINEERS SCHOLARSHIP**
Society of Flight Test Engineers (SFTE)
Contact: Alan R. Lawless, Director, Scholarship Program or
Allan Web, President, SFTE
PO Box 4047
Lancaster, CA 93539
Tel. (805) 538-9715
Fax: (805) 538-9715
Website: www.SFTE.ORG
E-mail: SFTE@Hughes.net

Deadline: Contact Sponsor

Eligibility: For members of SFTE, or children of SFTE members only

Award(s) & Amount(S): Varies, Range $200 - $2,000

❐ **SOUTHEASTERN AIRPORT MANAGERS ASSOCIATION (SAMA)**
Sam McKenzie
P.O. Box 35005
Greensboro, N.C. 27425
Tel. (910) 665-5600

Deadline: Contact Sponsor

Eligibility: There is one scholarship given to individual at each of the following universities: Embry-Riddle Aeronautical University, Auburn University, Louisiana Tech University, Middle Tennessee State University, and Delta State University. Institution must be located in the SAMA region and should have an approved four-year program directly related to airport management.

Award(s) & Amount(s): 5, $1,500

❐ **VIRGINIA AIRPORT OPERATORS COUNCIL AVIATION SCHOLARSHIP AWARD**
Virginia Airport Operators Council
Betty Wilson
5701 Huntsman Road
Sandston, VA 23150
Tel. (800) 292-1034

Deadline: February 16 each year

Eligibility: High school Senior with a "B" average or better planning a career in aviation who has been accepted and is enrolled in an accredited college in an aviation-related program. Limited to Virginia residents.

Award(s) & Amount(s): 1, $2,000

WOMEN

❐ **AMELIA EARHART RESEARCH SCHOLAR GRANT**
Ninety-Nines, Inc.
International Women Pilots
Will Rogers World Airport
P.O. Box 59965

Oklahoma City, OK 73159
Tel. (405) 685-7969

Deadline: December 31

Purpose: To provide financial assistance to a highly specialized professional scholar to work in her field of expertise in order to expand knowledge about women in aviation and space.

Eligibility: Highly specialized professional scholars are eligible to apply if they are interested in researching the role of women in aviation and space.

Award(s) & Amount(s): 1 each granting period. The amount awarded varies; generally, the award is at least $1,000.

❒ THE DOROTHY PENNEY SPACE CAMP SCHOLARSHIP
Contact: Sara Carson
Tel. (803) 576 7336

Eligibility: Contact Sponsor

Award(s) & Amount(s): Contact Sponsor

❒ ESTHER COMBES VANCE / VEIN VINE MEMORIAL FLIGHT TRAINING SCHOLARSHIP
The Montana Chapter of the 99s
Contact: Gail Sanchez-Eaton (406) 586 4126; or Tina Pomeroy (406) 222-6826 after 5:00 P.M.
1811 Baxter Drive
Bozeman, MT 59715

Deadline: February

Eligibility: The applicant rnay hold a student pilot's certificate, no other ratings or certificates may be held. The scholarship is open to any Montana female student pilot or a Montana female interested in leaming how to fly.

She must be a resident of the state of Montana and receive her flight training in Montana by a Montana registered CFI. Flight training must begin within three (3) months of receiving the award and be completed within two (2) years of receiving the award. If this timeline cannot be met, the scholarship will be forfeited. The Scholarship winner will be required to join the Montana Ninety Nines for a period of not less than two years.

Note: Payment will be made upon receipt of documentation of hours flown, payable to the recipient and the flight instructor or flight school. The documentation must be signed by both the recipient and the instructor. This must be submitted to the Montana Ninety Nines treasurer and payment will be made upon receipt. Requests for applications should be sent to the above address, please enclose a self addressed stamped envelope.

Award(s) & Amount(s): $500

❒ HELENE OVERLY SCHOLARSHIP
Women's Transport Seminar
Chicago Chapter
Scholarship Committee
P.O. Box 804535
Chicago, IL 60680

Deadline: February

Eligibilty: Contact Sponsor

Award(s) & Amount(s): $1500

❒ **INTERNATIONAL SOCIETY OF WOMEN AIRLINE PILOTS (ISA +21)**
Contact: Luan Meredith-Ward
2250 E. Tropicana Ave., Suite 19-395
Las Vegas, NV 89119-6594
Tel. (314) 845-7282
Website: http://www.iswap.org

Deadline: April 1, postmarked

❒ **The ISA International Career Scholarship**
ISA has earnmarked funds for a pilot who has already demonstrated her decision to join ISA members in this great profession. These scholarships are to be used for advanced pilot ratings such as the US FAA ATP certificate or equivalent.

Applicants who do not meet those requirements may pursue a **MERIT** scholarship. When funds are available, other unique scholarships are awarded, such as the **Fiorenza de Bernardi Merit Award,** and the **Holly Mullions Memorial Scholarship** (reserved for single mother). These merit scholarships will aid pilots endeavoring to fill some of the basic positions, i.e. CFI, CFII, MEI, or any of the equivalents.

❒ **The ISA International Airline Scholarship**
In recent years, the ISA scholarship program has grown due to the generosity of several airlines. ISA+21 has been able to offer aircraft type ratings (727, 737, 747, 757, and DC-10 aircraft) as well as flight engineer certificates to women actively pursuing an airline career.

Eligibility: Any woman pilot who is pursuing a career as a professional pilot, is able to demonstrate financial need, and meets the following requirements: All applicants must have a U.S. FAA Commercial Pilot Certificate with an Instrument Rating and a First Class Medical (or equivalent). For Merit Scholarship and Career Scholarship - a minimum of 750 flight hours. For a Flight Engineer Certificate Scholarship - minimum of 1,000 hours flight time and a current FE written. For a Type Rating Scholarship - an ATP Certificate and a current FE written.

Award(s) & Amount(s): Contact Sponsor

❒ **MARION BARNICK MEMORIAL SCHOLARSHIP**
Contact: Ann Tapey, Chairperson
Marion Barnick Scholarship Committee
860 S. Blaney Ave
Cupertino, CA 95014

Eligibility: Scholarships available to female private pilot member of the Ninety Nines and a student at San Jose State, Gavilian College, Foothill College or West Valley College in California.

Award(s) & Amount(s): $1,000

❒ **NATIONAL COUNCIL FOR WOMEN IN AVIATION/AEROSPACE**
PO Box 716
Lemont, IL 60439 - 0716
Tel. (800) 727 - NCWA (6292)
Fax: (630) 243 - 1828

Deadline: August

Description: The NCWA Scholarship Program assists women to attain career goals through continued education in aviation and aerospace

Eligibility: Current member in the NCWA. Selection is based on the completed application including the recommendation letters, financial need, and commitment to furthering your education in aviation and aerospace.

Award(s) & Amount(s): 3, $1,000

❒ PAM VAN DER LINDEN MEMORIAL SCHOLARSHIP
The Ninety-Nines. Inc.
Coyote Country Chapter
Penny Fedorchak, Chairman
241 Fox Fire Lane
Fallbrook, CA, 92028
Tel. (760) 728-0658
E-mail: JFedorchak@aol.com

Deadline: August

Eligibility: The applicant must be a female with at least a private pilot certificate. Limitations and Restrictions: Within one year of winning this scholarship the recipient MUST utilize the funds at a flight school approved by Coyote Country Chapter of the Ninety-Nines, Inc. Applicants should reside in the SouthWestern Section of the 99's and must attend a flight school located in: San Diego, Riverside, Imperial, San Bernardino, Los Angeles, Orange, Santa Barbara, or Ventura Counties.

Note: The recipient shall designate a flight school of their choice, subject to the restrictions above. The Coyote Country Chapter of the Ninety-Nines, Ins, will establish an account in the recipients name at the designated flight school. As bills are incurred the Coyote Country Chapter will pay all flight school expenses directly to the flight school.

Award(s) & Amount(s): 1, $750.00

❒ PRITCHARD CORPORATE AIR SERVICE INC. LEND A HAND SCHOLARSHIP
Tel. (602) 263-0190

Deadline: Contact Sponsor

Description: This scholarship is awarded to a certified female pilot to obtain a commercial rotor craft helicopter additional rating, and certified flight instructor rating

Eligibility: This is an extensive, eighteen-month program that anticipates all training and ratings to be completed in one year, followed by a minimum of six months employment as a flight instructor for Pritchard Air Service Program requires relocation to Novato, California for the 18 month period.

Award(s) & Amount(s): Contact Sponsor

❒ SAN FERNANDO VALLEY 99S CAREER SCHOLARSHIP
SFV 99s Career Scholarships
P.O. Box 8160
Van Nuys, CA 91409
Tel. (818) 989 0081

Deadline: Contact Sponsor

Description: Three scholarships available to men and women who want to pursue careers in Aviation. **Note:** 18 years old and live in greater Los Angeles Area

Award(s) & Amount(s): $1,000

❒ SAN FERNANDO VALLEY 99S FUTURE WOMEN PILOTS ASSISTANCE AWARDS
Future Women Pilot Program
P.O. Box 7142
Van Nuys, CA 91409
Tel. (818) 989 0081

Deadline: Contact Sponsor

Description: These are designed to assist women in obtaining their Private Pilot Certificates. **Note:** 18 years old and must reside in the greater Los Angeles Area

Award(s) & Amount(s): $400

❒ THE SOCIETY OF WOMEN ENGINEERS SCHOLARSHIP PROGRAM
Society of Women Engineers
Contact: Anette Sauer, Program Coordinator
120 Wall Street, 11th Floor
New York, NY 101)05-3902
Tel: (212) 51)9-9577
Fax (212) 509-0224
Website: http://www.swe.org
E-mail: hq@swe.org

Deadline: Varies

Description: As part of its national educational activities, Society of Women Engineers (SWE) administers over 90 scholarships annually, varying in amounts. All grants arc announced in the spring or summer for use during the following academic year. Grant payments are made in the fall for corporate-sponsored scholarships and in both the fall and spring for endowed scholarships, upon proof of registration. Application forms can be obtained through the Deans of Engineering at eligible schools, through SWE sections, Student sections and from SWE Headquarters. Requests to SWE Headquarters must be accompanied by a self-addressed stamped envelope.

Eligibility: All SWE scholarships are open only to women majoring in engineering or computer science in a college, or university with an ABET-accredited program or in a SWE-approved school and who will be in the specified year of study during the academic year the grant payment is made. Applicants For sophomore, junior, senior and graduate scholarships must have a grade point average of 3.5/4.0 or above. Additional specific requirements for the various scholarships are described in the following sections. Applicants will be considered for all scholarships for which they are eligible.

<u>Freshman Scholarships</u>

Deadline: May 15

The following scholarships are available only to incoming freshman.

❒ The Admiral Grace Murray Hopper Scholarships
Description: These scholarships were established in 1992 in memory of the "mother of computerized data automation in the naval service' for women entering the study of computer engineering or computer science in any form of a four-year program.

Award(s) & Amount(s): 5, $1000

❒ Anne Maureen Whitney Barrow Memorial Scholarship
Description: The scholarship is awarded to an engineering student. It is renewable for three additional years and, consequently, is awarded only once every four years.

Award(s) & Amount(s): 1, $5000

❒ Chrysler Corporation Scholarships
Description: These are awarded to students majoring in engineering or computer science.

Award(s) & Amount(s): 2, $1500

❒ General Electric Fund Scholarship
Description: They are renewable for three years upon evidence of continued academic achievement. In addition, the General Electric Fund provides a $500 travel grant for each entering freshman recipient to attend the SWE National Convention and Student Conference. Recipients must be U.S. citizens enrolled in engineering programs.

Award(s) & Amount(s): 3, $1000

❒ **The Lockheed Martin Corporation Scholarships**
Description: The scholarship IS awarded to engineering majors.

Award(s) & Amount(s): 2, $3000

❒ **The Nalco Foundation Scholarship**
Description: It is awarded to a chemical engineering major. It is renewable for three additional years and, consequently, is awarded only Once every four years.

Award(s) & Amount(s): 1, $1050

❒ **The Northrop Grumman Scholarships**
Description: These are awarded to students majoring in engineering or computer science.

Award(s) & Amount(s): 1, $1500; 2, $1000

❒ **The 3m Company Scholarships**
Description: These awards are presented to students entering engineering. Preference is given to chemical, electrical, industrial and mechanical engineering majors. Recipients must be U.S. citizens.

Award(s) & Amount(s): 3, $1050

❒ **The Trw Foundation Scholarships**
Description: These scholarships are administered on a local level by SWE's Best National, Regional and New Student Sections.

Award(s) & Amount(s): 1, $500; 10, $200

❒ **The Westinghouse Ibertha Lamme Scholarships**
Description: These scholarships were instituted to attract women to the field of engineering and are supported by the Westinghouse Educational Foundation. Recipients must be U.S. citizens enrolled in engineering programs.

Award(s) & Amount(s): 3, $1000

SOPHOMORE & JUNIOR AND SENIOR SCHOLARSHIPS

Deadline: February 1

The following scholarships are available only to students continuing in an engineering or computer science degree program. Completed applications including all supporting materials, must be postmarked no later than Recipients will be notified in May.

❒ **The Central Intelligence Agency Scholarship**
Description: These scholarships are awarded to an entering sophomore student majoring in electrical engineering or computer science. Recipients must be U.S. citizens.

Award(s) & Amount(s): 1, $1000

❒ **The Chrysler Corporation Scholarship**
Description: It is awarded to an entering sophomore junior, or senior woman who is a member of an underrepresented group in the engineering or computer science field.

Award(s) & Amount(s): 1, $1750

❒ **David Sarnoff Research Center Scholarship**
Description: Award to a woman engineering or computer science student, who is a U.S. citizen entering her junior year of study.

Award(s) & Amount(s): 1, $1500

❒ **Dorothy Lemke Howarth Scholarships**
Description: Scholarships are awarded to women engineering or computer science students entering their sophomore year who are U.S. citizens.

Award(s) & Amount(s): 5, $2000

❒ **General Motors Foundation Scholarships**
Description: Scholarships are awarded to students entering their junior year with a declared major in one of the following disciplines: mechanical, electrical, chemical industrial, materials, automotive or manufacturing engineering or engineering technology. Winners must demonstrate leadership by holding a position of responsibility in a student organization and exhibit career interest in the automotive industry or manufacturing environment. The scholarships are renewable for the senior year. In addition, the General Motors Foundation provides a $500 travel grant for each new recipient to attend the SWE National Convention and Student Conference.

Award(s) & Amount(s): 2, $1500

❒ **Ivy Parker Memorial Scholarship**
Description: The scholarship is presented to an engineering or computer science student entering her junior or senior year with a demonstrated need of financial assistance.

Award(s) & Amount(s): 1, $2000

❒ **Judith Resnik Memorial Scholarship**
Description: The award is presented to a student entering her senior year majoring in aerospace, aeronautical or astronautical engineering, who is also an active SWE Student Member.

Award(s) & Amount(s): 1, $2000

❒ **Lillian Moller Gilbreth Scholarship**
Description: The award is made to an engineering or computer science student of outstanding potential and achievement entering her junior or senior year.

Award(s) & Amount(s): 1, $5000

❒ **Lockheed Martin Fort Worth Scholarships**
Description: The awards are presented to entering juniors majoring in electrical or mechanical engineering.

Award(s) & Amount(s): 2, $1000

❒ **Maswe Memorial Scholarships**
Description: The awards are presented to engineering or computer science students entering their sophomore, junior or senior years and who demonstrate outstanding scholarship and financial need.

Award(s) & Amount(s): 2, $2000, 1, $1000

❒ **Northrop Corporation Founders Scholarship**
Description: First awarded in 1983 as an encouragement to students pursuing engineering degrees. The award is limited to a current SWE Student Member and a U.S. citizen entering her sophomore year.

Award(s) & Amount(s): 1, $1000

❒ **Rockwell Corporation Scholarship**
Description: These scholarships are awarded to minority women engineering or computer science students entering their junior years. The recipients must have demonstrated leadership ability.

Award(s) & Amount(s): 2, $3000

❒ **Stone And Webster Scholarships**
Description: Scholarships are presented to engineering or computer science students entering their sophomore, junior, or senior years.

Award(s) & Amount(s): 1, $1500; 3, $1000

❒ **United Technologies Corporation Scholarships**
Description: Established to encourage women engineering students to continue their education. Two awards are presented each year and are renewable for two years with continued academic achievement. The recipients must be U.S. citizens entering their sophomore year and majoring in electrical or mechanical engineering.

Award(s) & Amount(s): 2, $1000

GRADUATE SCHOLARSHIPs

Deadline: February 1

The following scholarships are available to students pursuing an advanced degree program. They are available only to students entering the first year of a master's program.

❒ **General Motors Foundation Graduate Scholarship**
Description: Requirements of discipline, leadership ability, and career interest are the same as those for the undergraduate General Motors Foundation Scholarships. In addition to the scholarship award, the General Motors Foundation provides a $500 travel grant for each recipient to attend the SWE National Convention and Student Conference.

Award(s) & Amount(s): 1, $1500

REENTRY SCHOLARSHIPS

Deadline: May 15

The reentry scholarships were established to assist women in obtaining the credentials necessary to reenter the job market as engineers. Eligibility is restricted to women who have been out of the engineering job market as well as out of school for a minimum of two years.

❒ **B.K. Krenzer Memorial Reentry Scholarship**
Description: Preference is given to degree engineers desiring to return to the workforce following a period of temporary retirement. Recipients may be entering any year of an engineering program, undergraduate or graduate, as full or part time students.

Award(s) & Amount(s): 1, $1000

❒ **Chrysler Corporation Reentry Scholarship**
Description: This scholarship is presented to an engineering or computer science student.

Award(s) & Amount(s): 1, $2000

❒ **Olive Lynn Salembier Scholarship**
Description: Recipients may be entering any undergraduate or graduate year, including doctoral programs, as full or part time students.

Award(s) & Amount(s): 1, $2000

❒ **THE TWEET COLEMAN AVIATION SCHOLARSHIP**
American Association of University Women
Contact: Jane Vierra
1802 Keeaumoku Street
Honolulu, HI 96822
Tel. (?) 537-4702

Deadline: February

Description: To establish and encourage professional growth in women seeking to enter the aviation field. To provided financial assistance to the aspiring Aviatrix.

Eligibility: **The applicant must be a female resident of Hawaii**; a college graduate or currently attending an accredited college in the State of Hawaii; and able to pass a 1st Class FAA medical exam.

Award(s) and Amount(s): Contact Sponsor

❒ **STUDENT PILOT SCHOLARSHIPS**
Contact: Trine Jorgensen
Tel. (303) 989 1948

Eligibility: Colorado female student pilot in financial need who has passed the written.

Award(s) & Amount(s): $500

❒ WHIRLY-GIRLS SCHOLARSHIPS
The Men's Auxiliaries of the Whirly Girls
The Whirly-Girls Scholarship Fund, Inc.
Contact: Lee Hixon, President
3674 Andreas Hills Dr.
Palm Springs, CA 92264
Tel. (760) 325-9250
Fax (760) 325-8099
E-mail: aleeh426@aol.com

Deadline: October 31

Eligibility: **Applicants must be female pilot holding at least a Private Pilot license or higher in fixed wing.** The scholarship is to be used toward a transition helicopter rating. The scholarship is for rotorwing training at a school approved by the Whirly Girls..

Award(s) & Amount(s): $4,500

❒ WOMEN IN AVIATION, INTERNATIONAL
Contact: Peggy Baty
Morningstar Airport
3647 S.R. 503 S.
W. Alexandria, OH 45381
Tel. (937) 839-4647
Fax: (937) 839-4645

Deadline: March

Several aviation companies have coordinated efforts with Women in Aviation, International to provide numerous scholarships to be awarded at their March 1999 conference in Dallas, Texas. To receive an application form(s) for any of the listed scholarships call or write the address listed. The list below are examples of some scholarships awarded by the Women In Aviation.

GENERAL SCHOLARSHIPS

❒ Airbus Leadership Grant
Description: The scholarship will be awarded to a woman at the level of sophomore year or above who is pursuing a college degree in an aviation-related field, who has achieved a minimum of a 2.0/4.0 GPA. and who has exhibited leadership potential. All applicants should submit a 350-word essay which addresses their career aspirations and explains how they have exhibited leadership skills.

Award(s) & Amount(s): $1,000

❒ Amelia Earhart Society Career Enhancement Scholarship
Description: The scholarship provided by ABS will be awarded to a woman who is active in aerospace and is seeking financial support to enhance her aerospace career. Applicants may be a full-time or part-time employee in the aerospace industry who has demonstrated commitment as a community supporter or care-giver. All applicants should submit a 350-word essay which addresses their career aspirations and goals. In addition, they should explain how they have exhibited personal growth through community service and care giving.

Award(s) & Amount(s): $500
❒ Boeing Company Career Enhancement Scholarship
Description: The Boeing Company will award a scholarship to a woman who wishes to advance her career in an

aerospace technology or a related management field. Applicants may be full-time or part-time employees currently in the aerospace industry or related field. Also eligible are students pursuing an aviation related degree who are at the Junior level with a GPA. of 2.5/4.0.

All applicants should submit a 350-word essay which addresses their career aspirations and education goals. In addition, they should explain how they have exhibited career growth and development, as appropriate.

Award(s) & Amount(s): $1,000

❒ **Colleen Barrett Aviation Management Scholarship**
Description: A scholarship was established in honor of Colleen Barrett, Vice President of Customers, Southwest Airlines. The scholarship will be awarded to a woman in an aviation management field who has exemplified the traits of leadership, community spirit and volunteerism. The scholarship can be used to attend a leadership-related course or seminar or other means of raising the individual's level of management position.

Applicants should include a cover letter describing their current job position, community involvement and volunteer work and how the scholarship money will be used. A one page resume depicting the applicant's previous work history along is also desirable.

Award(s) & Amount(s): $750

❒ ***Flight Training* Magazine Scholarships**
Description: *Flight Training* magazine is offering the opportunity for two students to attend the 1999 International Women in Aviation Conference in Orlando, FL. In return for paying for travel expenses, the students will assist in the *Flight Training* exhibit booth. The scholarships are available to full time students pursuing an aviation career. The recipients must be able to travel to Orlando and attend the entire Conference, March 18-20. Scholarship includes airfare, hotel (double occupancy) and meals during the conference. Conference registration will be donated by Women in Aviation, International.

Applications must be received by ***Flight Training*** no later than Friday, January 8, 1999. Send a one-page letter describing yourself and why you want to attend the Conference. In addition, you are required to write a one-page double-spaced, typewritten essay on either "Marketing General Aviation to the General Public" or "How to Attract More Women to Flying." These two pages only should be mailed to: Women in Aviation Scholarship, ***Flight Training*** Magazine, 201 Main Street, Parkville, MO 64152.

Award(s) & Amount(s): $ 1,000/each. (Total value - $2,000)

❒ **Women in Aviation. International Achievement Awards**
Description: Scholarships will be awarded to full-time college or university students pursuing any type of aviation or aviation-related career. A third $500 scholarship will be awarded to an individual, not required to be a student, pursuing any type of aviation career.

Award(s) & Amount(s): 2, $500

FLIGHT SCHOLARSHIPS

❒ **Airbus A320 Type Rating Certificate Scholarships**
Description: The requirements for this scholarship include a bachelor's degree, commercial pilot certificate, instrument rating, certified flight instructor certificate and multi-engine rating. Preference will be given to those candidates who are graduates of Spectrum-type programs, featuring technical flight management systems and glass cockpit training. Two scholarships will be awarded.

Award(s) & Amount(s): 2, $30,000/each. (Total value - $60,000)

❒ **American Airlines B-727 Flight Training Award**
Description: Any woman who is pursuing a career as a professional pilot, who is able to demonstrate financial

need and who meets the prerequisites may apply for this award. These recipients will be given the standard AA initial aircraft training program. Two scholarships will be awarded. The prerequisites are the same as those defined in the Northwest Airlines scholarship.

Award(s) & Amount(s): 2, $ 18,000/each. (Total value - $36,000)

❒ **Cessna Aircraft Company Private Pilot Scholarship**
Description: Cessna Aircraft Company is offering a scholarship to earn a private pilot's certificate. The applicant must be able to pass a flight physical, have no prior logbook entry flight training, and willing to complete the course at a Cessna Pilot Center within a 12-month period. The all-new interactive Computer Based instruction kit ($200 value) will be provided along with a series of credits which will be paid directly to the flight school. This will include $500 after medical certificate is received; $500 upon completion of first solo flight; $1000 after completion of first cross-country flight; and balance, up to $1,500, when the FAA written exam is passed and the private pilot certificate is received. Please include in the 350-word essay why you want to learn to fly and how you intend to use the certificate once obtained.

Award(s) & Amount(s): 1, $3,700

❒ **Jeppesen Sanderson Company Private Pilot Scholarship**
Description: Jeppesen Sanderson Company is providing $1,000 scholarships to assist two individuals beginning work on their Private Pilot - airplane certificate. This scholarship has no age requirement, other than the minimum required by the FAA, or education requirements.

Award(s) & Amount(s): 2, $1,000/each. (Total value - $2,000)

❒ **Northwest Airlines Type Rating Award**
Description: Any woman who is pursuing a career as a professional pilot, who is able to demonstrate financial need and who meets the following prerequisites may apply for this type rating award. (Previous recipients have received DC-9 and B-747400 type ratings.)

For Captain: United States applicants must have a commercial pilot certificate with an instrument rating, a minimum of 1500 hours total time, including 800 hours multi-engine time and a current first class FAA medical certificate.

For Flight Engineer: United States applicants must have a commercial pilot certificate with an instrument rating, a minimum of 1000 hours total time and a current first class FAA medical certificate.

International applicants for either position must meet the above requirements. In addition, you must include your country's pilot hiring requirements and the equivalent of a current first class FAA medical certificate.

Award(s) & Amount(s): $40,000 to $50,000

❒ **SimuFlite Citation II Corporate Aircraft Training Scholarship**
Description: SimuFlite Training International will award a corporate aircraft training scholarship: It will include Citation II initial training resulting in a type rating upon successful completion of the course. The applicant is responsible for all travel and housing related costs for the 13-day course.

Criteria includes, in addition to the general scholarship requirements, a current first class medical, commercial pilot certificate with multi-engine and instrument ratings and a certified flight instructor certificate with instrument airplane endorsement. Applicants must have flown a minimum of 50 hours PIC or SIC within the previous 12 months. Applicants must be available for training during June or July 1999. Applicants must demonstrate an interest in corporate or business aviation by writing a 350-word essay on how this scholarship will fit into their plans to enter corporate/business aviation. This essay must demonstrate how the applicant will use the SimuFlite Scholarship to pursue a flight-related career.

Award(s) & Amount(s): 1, $10,500

❒ **United Airlines Type Rating Scholarship**
Description: Potential candidates for UAL Type Rating Scholarships may not hold any previous large transport category aircraft type ratings on their pilot certificate. Eligibility is limited to potential candidates who are not currently flying with a regional or national airline. Second class FAA medical certificate with the ability to obtain a first class certificate is required. One must have obtained their commercial certificate with a minimum of 350 hours of flight experience in fixed wing aircraft as command pilot or co-pilot (not including student time). Applicant must be a high school graduate. Applicant must be able to arrange schedule to attend the five (5) to six (6) week training program as well as support yourself during this period (including housing and meals).

Award(s) & Amount(s): 2, $10,000/each. (Total value - $20,000)

MAINTENANCE SCHOLARSHIPS

❒ **Aircraft Electronics Association Aviation Maintenance Scholarship**
Description: The Aircraft Electronics Association is offering a scholarship for a female student seeking a degree in the aviation maintenance field (preference will be given to an avionics major) at an accredited college or technical school offering such a degree. A 2.75/4.00 GPA. is required

Award(s) & Amount(s): 1, $1,000

❒ **ATP Maintenance Technician of the Year Award**
Description: Aircraft Technical Publishers awards an annual Maintenance Technician of the Year Award to an individual. The recipient receives a plaque and $1,000. Qualifications for the award include being a licensed A&P or IA, recommendation letters from colleagues or employers and a minimum of three years experience in an aviation maintenance field. A 350-word descriptive essay should include achievements, attitude toward self and others, dedication to career and demonstration of professionalism.

Award(s) & Amount(s): 1, $1,000

❒ **Bombardier Challenger Initial Maintenance Scholarship**
Description: Bombardier is offering a free Challenger Initial Maintenance Course to an individual. The four-week course requires students to have previous aircraft experience, other jet experience recommended. It is designed to provide an introduction to the Challenger aircraft and its systems. The student will gain a thorough understanding of the operational features and maintenance requirements of the aircraft.

Award(s) & Amount(s): 1, $8,400

❒ **SimuFlite Maintenance Scholarship**
Description: SimuFlite Training International will award a Citation III maintenance initial training scholarship. The scholarship is designed for technicians who have minimal experience working on the aircraft type. The applicant is responsible for all housing and travel costs. It is a 10-day course including 60 hours of instruction. This maintenance scholarship is intended for those currently working in the maintenance field or attending a college, university or vocational school.

Award(s) & Amount(s): 1, $3,800

❒ **WOMEN'S TRANSPORT SEMINAR UNDERGRADUATE SCHOLARSHIP**
Women's Transport Seminar

C/O Barton-Aschman Associates, Inc.
820 Davis St., Suite 300
Evanston, IL 60201

Deadline: February

Eligibility: Contact Sponsor

Award(s) & Amount(s): $2000

❒ WOMEN MILITARY AVIATORS, INC. SCHOLARSHIP
Women Military Aviators, Inc.
Contact: Major Carla Gammon
2605 Pro Tour Drive
Belleville, IL 62220

Deadline: June

Description: Women Military Aviators, Inc. offers a scholarship for woman interested in a career in aviation. The scholarship is dedicated to the Women Airforce Service Pilots (WASPs). WASPs sacrifices and pioneer spirit continue to be an inspiration to women in aviation today. Funding will be provided to pursue an FAA private pilot or advanced rating.

Eligibility: Selections will be based primarily on the applicant's goals and aspirations. Ambition to further advance women in aviation. Financial Need.

Award(s) & Amount(s): 1, $2,500

❒ WOMEN OF THE NATIONAL AGRICULTURAL AVIATION ASSOCIATION
NAAA
1005 E Street SE
Washington D.C. 20003
Tel.: (202) 546-5722
Fax: (202) 546-5726

Deadline: Contact Sponsor

Eligibility: The competition is open to the relatives of National Agricultural Aviation Association members, who have paid their NAAA dues on or before June 15 of the current year. All entrants must be students enrolled in continuing education. For more specific information, contact the NAAA.

Award(s) & Amount(s): $1,500

❒ WOMEN SOARING PILOTS' ASSOCIATION (WSPA)
Contact Pat Valdata
E-mail: pvaldata@prodigy.net.

Deadline: Contact Sponsor

Description: Two scholarships are being offered by the Women Soaring Pilots' Association (WSPA) for women wanting to further their soaring education. For further information.

Eligibility: Contact Sponsor

Award(s) & Amount(s): Contact Sponsor

LOAN

This section provides you with a listing of organizations and institutions offering loans to students. Funding for college and flight training represents a significant hurdle for most students. Students rely on loans from family and lending institutions to raise the necessary tuition and flight training dollars.

❐ ARMY AVIATION ASSOCIATION OF AMERICA SCHOLARSHIP FOUNDATION
49 Richmondville Avenue
Westport, CT 06880-2000
Tel. (203) 226-8184
Fax: (203) 222-9863

Deadline: April

Purpose: To provide funding, in the form of loans, for the postsecondary education of members of the Army Aviation Association of America (AAAA) or of their dependents.

Eligibility: This program is open to members of the AAAA or to their spouses, siblings, or children. Special consideration is given to applications submitted or sponsored by warrant officers and enlisted personnel.

Award(s) & Amount(s): At least 5 each year. The amount loaned varies but is generally at least $1,000 per year. All loans are interest free. Duration: Up to 4 years.

❐ ACADEMIC MANAGEMENT SERVICES (AMS)
Tel. (800) 635-0120
Website: http://www.amsweb.com

Description: AMS offers subsidized and unsubsidized financial assistance to students attending participating schools. Contact AMS to see whether the school you are interested in is covered. Qualifications are based on good credit and sufficient income.

❐ AIRMEN MEMORIAL FOUNDATION FEDERAL STUDENT LOAN PROGRAM IN COOPERATION WITH SIGNET BANK
Educational Funding Services
PO Box 1573
Baltimore, MD 21203 - 1573
Tel. (800) 955-0005

Description: Three types of federally sponsored student loans are available through The Airmen memorial Foundation. These loans must be used to pay for education related expenses such as tuition, books, room and board, and other living expenses. All offer attractive interest rates and convenient repayment terms. Call their student loan processor signet bank at the number listed above, say that AMF referred you, and ask for a free student loan application kit.

❐ AOPA'S FLIGHT FUND
Tel. (800) 441-7048 ext. 65932 or 82058

Description: AOPA's flight fund provides eligible flight students with a revolving line of credit of up to $25,000. Students must meet normal credit requirements (minimum gross annual income and an established credit history) and be an AOPA members. Loan applications may be completed by telephone (800-441-7048 ext. 65932) of Fax (302-457-3111) with a decision in one hour. Interest Rate 14.0%.

❒ **AVCO FINANCIAL SERVICES**
Tel. (800) 835-2826

Description: AVCO offers two types of loans: revolving lines of credit and installment loans. Annual interest rate 13.99 - 17.99%. Qualifications are based on good credit and sufficient income. Payment plans start 30 days from receiving the funds (not deferred). Repayment terms: up to 5 years

---

❒ **CAREER EDUCATION LOAN PROGRAM (CEL)**
Tel. (800) 244-8750

Description: CEL is a private, credit-based loan program for students enrolled in approved, eligible institutions. Call CEL to see if the school you are interested in is covered. Under the CEL program, you or any member of your family can borrow up to $25,000 a year. CEL features competitive rates, no application fee, a 15 year repayment term and no prepayment penalty. Qualifications are based on good credit and sufficient income.

❒ **THE EDUCATION RESOURCES INSTITUTE (TERI)**
Tel. (800) 255-TERI
Website: www.teri.org

Description: TERI offers a variety of loan programs to students attending participating and/or accredited schools. Contact TERI to see whether the school you are interested in is covered. Qualifications are based on good credit and sufficient income.

❒ **EDUCATION CREDIT CORPORATION**
Tel. (800) 627-5001

Description: Aviation Credit Line; Loan amount: $3,000 to $35,000; Interest rate: 12 - 12.5%; Co-signors allowed; Repayment begins 30 days after approval, no prepayment penalty; Repayment terms: up to 5 years.

❒ **FEDERAL EMPLOYEES EDUCATION ASSOCIATION (FEEA)**
8441 W. Bowles , Suite 200
Littleton, CO 80123-3245
Tel. (800) 323-4140

Description: FEEA has scholarships and educational loans available to federal civilian employees and their dependents. The scholarships are based on academic merit. Applications may be obtained by writing FEEA and including a self-addressed, postage-paid (.32) # 10 envelope.

❒ **HATTIE M. STRONG FOUNDATION STUDENT LOANS**
Contact: Hattie M. Strong Foundation
1735 Eye St. NW #705
Washington D.C. 20006

Eligibility: Interest free loans for students entering their final year of study at four year college.

Award(s) & Amount(s): $2,500

❒ **KNIGHTS TEMPLAR SPECIAL LOW INTEREST LOANS**
Knights Templar Education Foundation
Contact: Paul Rodenhauser
5096 W. Elston Ave. #A101
Chicago, IL 60630

Description: Loan program provides financial assistance for vocation, technical, or professional training and for junior-senior undergraduates and graduate students. Loan interest rate is 5% and must be repaid within four years after graduation.

Award(s) & Amount(s): $3,000

❐ **MBNA**
AOPA Gold Option Loan
Tel. (800) 847-7378

Description: Must be a member of AOPA ($39 per year); Loan amount: $5,000 to $25,000; Interest rate: 14.99%; Co-signors allowed; Repayment terms: up to 84 months for repayment, no prepayment penalty

---

❐ **NELLIE MAE**
Excell Famiily Education Loan
Tel. (800) 634-9308

Description: Loan amount: entire cost of program (including books, supplies, room and board, and expenses); Interest rate: Prime plus .75 - 2%; Co-signors: allowed; Repayment: begins 45 days after graduation' allowed to defer principal (pay interest only as a student); Repayment terms: up to 15 years

---

❐ **THE RETIRED OFFICERS ASSOCIATION**
201 North Washington St.
Alexandria, VA 22314
Tel. (703) 549-2311

Description: Provides interest-free loans to undergraduate students who are dependent children of active, reserve and retired uniformed service personnel.

Scholarship Deadlines

THIS PAGE INTENTIONALLY LEFT BLANK

SCHOLARSHIPS	DEADLINES
Some deadline dates varies from year to year. Contact the sponsor for the exact deadline date. **ALL MAJORS**	
AAAE Scholarship Program, The	Nofity
Air Traffic Control Association Scholarship	
✈ Half to Full-time Student Candidate	May 1
✈ Full-time Employee Candidate	May 1
✈ Children of Air Traffic Control Specialists	May 1
Allen H. and Nydia Meyers Foundation	March 15
Aviation Council of Pennsylvania	Nofity
Aviation Distributors and Manufacturers Association	March 15
Aviation Insurance Association Scholarship	September 30
Baccalaureate Degree Completion Program	Nofity
Bill Falck Memorial Scholarship	April
Charles H. Grant Scholarship	Nofity
Charlie Wells Memorial Aviation Scholarship	August
Charlotte Woods Memorial Scholarship (See Transport. Clubs Int'l Scholarship)	April
Denny Lydic Scholarship (See Transportation Clubs Int'l Scholarship)	April
Donald Burnside Memorial Scholarship	March 31
EAA Aviation Achievement Scholarships	April 1
Educational Communications Scholarships	Nofity
Eugene S. Kropf Scholarship	May
Ginger & Fred Deines Mexico Scholarship (See Transp. Clubs Int'l Scholarship)	April
Harold S. Wood Scholarship	February
Herbert L. Cox Memorial Scholarship	April 1
Hooper Memorial Scholarship (See Transportation Clubs Int'l Scholarship)	April
Landrum & Brown scholarship, The	Nofity
MAPA Safety Foundation, Inc.	September
Marine Corps Historical Center Research Grants	Nofity
McAllister Memorial Scholarship	March 1
MEA Aviation Scholarships	December 31
Montgomery GI Bill (Active Duty)	Nofity
Montgomery GI Bill (Selected Reserve)	Nofity
National Agricultural Aviation Association	Nofity
National Business Aircraft Association Scholarship	October 15
Pioneers of Flight Scholarship	November 15
Rick Leonard Memorial Scholarship, The	April 1
Safe Association Scholarship	June 1
Silver Dart Aviation History Award	March 15
Teledyne Continental Aviation Excellence Scholarship	April
Texas Transportation Scholarship (See Transportation Clubs Int'l Scholarship)	April
Transportation Clubs International Scholarships	April

USAIG PDP Scholarship	August
Wilfred M. Post Jr. Aviation Scholarship	March
AERONAUTICS / ASTRONAUTICS	
American Institute of Aeronautics & Astronautics Industry Scholarships (AIAA)	
✈ The Abe M. Zarem Award for Distinguished Achievement	January 31
AIAA Foundation Undergraduate Scholarship Program	January 15
AIAA Foundation Graduate Awards	
✈ AIAA Gordan C. Oats Air Breathing Propulsion Award	January 15
✈ AIAA Liquid Propulsion Award	January 15
✈ The Martin Summerfield Propellants and Combustion Graduate Award	January 15
✈ Open Topic	January 15
✈ AIAA William T. Piper, Sr. General Aviation	January 15
Allen H. Meyers Scholarship Foundation	March 15
American Society of Naval Engineering Program	February 15
Electronic Industries Foundation Scholarship	February 1
Silver Dart Aviation History Award	March 15
U.S. Airforce ROTC	Nofity
United States Space Foundation	Nofity
Vertical Flight Foundation Scholarships	Nofity
AEROSPACE SCIENCE	
Allen H. Meyers scholarship Foundation	March 15
Dr. Robert H. Goddard Space Science & Engineering Scholarship	January
I.T. Undergraduate Research Assistanceship Program	Nofity
Montana Space Grant Consortium	April
R. Minkin Aeropsace Engineering Scholarship	Nofity
U.S. Air Force ROTC	Nofity
Undergraduate / Graduate Engineering Scholarships	Nofity
ADMINISTRATION / MANAGEMENT	
AAAE Scholarship	May 15
ACI-NA Commissioners Roundtable Scholarship	November 1
ADMA Scholarship Program	March 15
Air Traffic Control Association Scholarship	Nofity
Boeing Student Research Award	February
Fred A. Hooper Memorial Scholarship, THe	March 31
National BusinessAviation Association Scholarship	Nofity
Southeastern Airport Managers Association Scholarship	Nofity
Wilfred M. Post, Jr. Aviation Scholarship, The	March 1
AVIATION MAINTENANCE	
ADMA Scholarship Program	March 15
Aviation Council of Pennsylvania Scholarship	July 31
Aviation Maintenance Educational Fund	Nofity
Career Quest Scholarship Program	October 15
Joseph Frasca Excellence in Aviation scholarship, The	April 15

Scholarship	Deadline
Mechanic / Technician Scholarship Award Program	September
PAMA Scholarship	Nofity
R. Minkin Aerospace Engineering Scholarship	Nofity
William M. Fanning Maintenance Scholarship	Nofity

AVIONICS / AIRCRAFT ELECTRONICS

Scholarship	Deadline
Aircraft Electronics Association Educational Foundation	March 1
✈ AEA Educational Foundation Pilot Training Scholarship	March 1
✈ Bose Corporation Avionics Scholarship	March 1
✈ Bud Glover Memorial Scholarship	March 1
✈ Castleberry Instruments Scholarship	March 1
✈ Chuck Peacock Honary Scholarship	March 1
✈ Colege of Aeronautics Scholarship	March 1
✈ David Arver Memorial Scholarship	March 1
✈ Dutch & Ginger Arver Scholarship	March 1
✈ EDMO Distribution Scholarship	March 1
✈ Gene Baker Memorial Scholarship	March 1
✈ Gulf Coast Avionics Scholarsdhip to Fox Valley Technical College	March 1
✈ Jim Cook Honary Scholarship, The	March 1
✈ Leon Harris/Les Nichols Memorial Scholarship to Spartan School of Aeronautics	March 1
✈ Lowell Glover Memorial Scholarship	March 1
✈ McCoy Avionics Scholarships	March 1
✈ Mid-Continent Instrument Scholarship	March 1
✈ Navair Limited Scholarship	March 1
✈ Northern Airborne Technology Scholarship	March 1
✈ Paul & Blanche Wulfsberg Scholarship	March 1
✈ Plane & Pilot Magazine/GARMIN Scholarship	March 1
✈ Robert Kimmerly Memorial Scholarship	March 1
✈ Russell Leroy Jones Memorial Scholarship to Colorado Aero Tech	March 1
✈ Terra Avionics Collegiate Scholarship	March 1
McCoy Avionics Scholarships	March 1
William E. Jackson Award	June 30

FELLOWSHIPS

Fellowship	Deadline
Amelia Earhart Fellowship Awards	November 1
Arizonia / NASA Space Grant Graduate Fellowship	Nofity
Frances Shaw Fellowship & Internship	February
National Air & Space Museum Fellowship	January 15
✈ A. Verville Fellowship	January 15
✈ Guffenheim Fellowship	January 15
✈ Ramsey Fellowship In naval Aviation History	January 15
Nebraska Space Grant Scholarships and Fellowships	Nofity
Tennessee Space Grant Consortium Graduate Fellowship Program	Nofity
Transportation Research Board Graduate Research Award Program VIII	Nofity
U.S. Air Force Dissertation Year Fellowship in U.S.	March 13

FLIGHT TRAINING

ADMA Scholarship Program	March 15
AERO Club of New England, The	
✈ New Advanced Pilot Scholarships	Nofity
✈ Ongoing Advanced Pilot Scholarships	Nofity
Amelia Earhart Memorial Scholarship	December 31
American Airlines First Officer Candidate Course Scholarship	May 1
American Helicopter Society	Nofity
Anne Marie Morrissey Aviation Scholarship	January 1
Aviation Council of Pennsylvania Scholarship	July 31
Cadet Youth Flight Scholarship	June 1
Comair Airline Pilot Scholarship	July
Dan L. Meisinger, Sr. Memorial Learn To Fly Scholarship	November
EAA Aviation Achievement Scholarships	April 1
Florenza De Banardi Merit Award	January 15
Franl P. Lahm, Flight 9 Scholarship	February
Gogos Scholarships Program	April 30
James R. Miresse Scholarship	Nofity
John E. Godwin, Jr. Memorial Scholarship	November
Joseph Frasca Excellence in Aviation Scholarship	April 15
Lt. Kara Hultgreen memorial Scholarship	May
Mary J. McGrath Scholarship	Nofity
Minnesota Aviation Trade Association Flght Scholarship	Nofity
Revolution Helicopter Corporation	Nofit
Richard Lee Vernon Aviation Scholarship, The	Nofity
SimuFlite Advanced Crew Training Scholarship	April 1
Soaring Society of America Youth Soaring Scholarship	Nofity
Sporty's Aviation Scholarship Program	January 15
U.S. Air Force ROTC	Nofity
Virginia Airport Operators Council Aviation Scholarship Award	February 16
Wagner Foundation Professional Pilot Scholarship, The	Nofity

GRANTS FOR EDUCATORS AND CHAPTERS

Aeropsace Education Foundation	Nofity
✈ Direct Grants to Educators	Nofity
✈ Direct Grants to AFJROTC Units & Civil Air Patrol Squadrons	Nofity
✈ Chapter Matching Grants for Aeropsace Education Programs	Nofity

MILITARY AFFILIATION

Aerospace Education Foundation	
✈ AEF Air Force Spouse Scholarship	Nofity
✈ Aeropsace Education Foundation Air Force Spouse Scholarship	Nofity
✈ AFROTC Angel Flight and Silver Wing Scholarship	Nofity
✈ Dr. Theodore von Karman Graduate Scholarship Program	Nofity
✈ Eagle Grant Enlisted Tuition Assistance	Nofity
✈ Janet R. (Wisemandle) Whittle Memorial Scholarship	Nofity

Air Force Aid Society	Nofity
Air Force Sergeant Association & Airmen Memorial Foundation Scholarship	Nofity
American Meteorology Society	Nofity
Association of Graduates of the U.S. Air Force Academy	Nofity
Budweiser USO Scholarship	March 1
Caption Jodi Callahan Memorial Scholarship	July
Daedalian Foundation	Nofity
Disabled American Veterans Auxiliary	April 15
National Women's Relief Corps	Nofity
Non-Commissioned Officers Association	Nofity
Reserved Officers Association	Nofity
Retired Enlisted Association, The	Nofity
Retired Officers Association	Nofity
Society of the Strategic Air Command	Nofity
Veteran's Adminstration	Nofity

SCHOOL - SPONSORED

College of Aeronautics	
For New Freshmen:	Nofity
✈ Community Scholarship Program	Nofity
✈ Founders' Scholarship Program	Nofity
✈ Freshman Academic Scholarships	Nofity
✈ Leon D. Star, M.D. Memorial Scholarship	Nofity
✈ Kiwanis Scholarship	Nofity
✈ Vocational Industrial Clubs of America (VICA) Scholarship	Nofity
New Transfer Students	Nofity
✈ Transfer Scholarships	Nofity
Continuing Students	Nofity
✈ Academic Scholarship	Nofity
✈ William Smart Alumni Association Scholarship	Nofity
✈ Board of Trustees Grants	Nofity
New York State Financial Aid Programs	Nofity
✈ Tuition Assistance Program (TAP)	Nofity
✈ Aid for Part Time Study (APTS)	Nofity
Delaware State University	Nofity
Eastern Kentucky University	Nofity
Florida Institute of Technology	Nofity
Fort Hays States University	Nofity
Guilford Technical Community College	December 1
Indiana State University Aviation Scholarship	
✈ Aviation Association of Indiana Scholarship	Nofity
✈ Capt. Ralph C. Miller Memorial Scholarship	Nofity
✈ Dennis J. Hunter Memorial Scholarship	Nofity
✈ Jeffrey Hardaway Memorial Scholarship	Nofity
✈ John A. Merritt Memorial Scholarship	Nofity
✈ Kenneth S. Papkoff Memorial Scholarship	Nofity
✈ Quentin R. Beecher Memorial Scholarship	Nofity
Iowa Central Community College	Nofity

Jacksonville University	February 1
Lewis University	Nofity
Miami - Dade Community College	March
Kansas State University	February
Middle Tennessee State University	
✈ Aerospace Graduate Scholarship	Nofity
✈ Airshow Aviation Scholarship	Nofity
✈ Colonel Jean Jack Aerospace Scholarship	Nofity
✈ Col. Harry E. Slater Memorial Scholarship	Nofity
✈ Dr. Wallace R. Maples Aerospace Scholarship	Nofity
✈ Don Ace Memorial Aerospace Scholarship	Nofity
✈ Frank & Harriett Hedrick Memorial Aviation Scholarship	Nofity
✈ Freshmen Scholarship to the Aerospace Program	Nofity
✈ H. Miller Lanier Memorial Aviation Scholarship	Nofity
✈ Jeffrey Clayton McCrudden Memorial Aviation Scholarship	Nofity
✈ Jim Price Jr. Memorial Scholarship	Nofity
✈ Metropolitan Nashville Airport Authority Aviation Scholarship	Nofity
✈ Southeastern Airport Managers Assos/AAAE Southeast	Nofity
Northwestern Michigan College	
✈ Aviation Division Scholarship	Nofity
✈ The Frank P. Macartney Foundation	Nofity
✈ Holts Claw Memorial Scholarship	Nofity
Oklahoma State University	March 31
Southern Illinois University at Carbondale	
✈ Cessna, Gatewood, and Staggerwing	April
✈ Jerry Kennedy Aviation Career Advancement Scholarship	October
✈ William R. Norwood Aviation Scholarship	October
Parks College of St. Louis University	Nofity
Texas Southern University	Nofity
University of Cincinnati, Clermont College	March
University of Illinois - Institute of Aviation	Nofity
University of Marylanf Eastern Shore	Nofity
University of Minnesota	
✈ Boeing Scholarship	Nofity
✈ I.T. Undergraduate Research Assistantship Program	Nofity
✈ R. Minkin Aerospace Engineering Scholarship	Nofity
Univeristy of Oklahoma - Aviation Department	Nofity
Winona State University	Nofity

SPECIAL INTEREST & AFFILIATION

AAAA Freshman Scholarships	April
AAAA Graduate Scholarships	April
AAAA Spouse Scholarships	April
AAAA Upperclassmen Scholarships	April
Airline Pilots Association Scholarship	April
Alpha Eta Rho Scholarship	March 9
American Geological Institute Minority Participation Program	Nofity
Aviation Scholarship, OBAP	Nofity
Captain Grant T. Donnell & Admiral Jimmy Thach Memorial	July

Scholarships	
Illinois Pilots Association Scholarship	April
NASA Undergraduate Student Researchers Program	Nofity
Navy Tuition Assistance Program	April
NHCFAE College Student Scholarships	April
Nebraska Space Grant Scholarships and Fellowships	Nofity
Oregon Educational Aid for Veterans	Nofity
PalWaukee Airport Pilots Association	June
Rhode Island Pilots Association	February 20
Robbins Airport Scholarship	April 15
Society of Flight Test Engineers Scholarship	Nofity
Southeastern Airport Managers Association (SAMA)	Nofity
Virginia Airport Operators Council Aviation Scholarship Award	February 16
WOMEN	
Amelia Earhart Research Scholar Grant	December 31
Dorothy Penney Space Camp Scholarship	Contact Sponsor
Esther Combes Vance / Vein Vine Memorial Flight Training Scholarship	February
Helene Overly Scholarship	February
Internatioanl Society of Women Airline Pilots (ISA+21)	
✈ The ISA International Career Scholarship	April
✈ The ISA International Airline Scholarship	April
Marion Barnick Memorial Scholarship	Nofity
National Council for Women in Aviation / Aerospace	August
Pam Van Der Linden Memorial Scholarship	August
Pritchard Corporate Air Service Inc. Lend A Hand Scholarship	Nofity
San Fernando Valley 99s Career Scholarship	Nofity
San Fernando Valley 99s Future Women Pilot Assistance Award	Nofity
Society of WomenEngineers Scholarship, The	Nofity
<u>Freshman Scholarships</u>	
✈ Admiral Grace Murray Hopper Scholarships	May 15
✈ Anne Maureen Whitney Barrow Memorial Scholarship	May 15
✈ Chrysler Corporation Scholarships	May 15
✈ General Electric Fund Scholarship	May 15
✈ Lockheed Martin Corporation Scholarships, The	May 15
✈ The Nalco Foundation Scholarship	May 15
✈ The Northrop Grumman Scholarships	May 15
✈ 3m Company Scholarships, The	May 15
✈ Trw Foundation Scholarships, The	May 15
✈ Westinghouse lbertha Lamme Scholarships, The	May 15
<u>Sophomore, Junior & Senior Scholarships</u>	
✈ Central Intelligence Agency Scholarship, The	February 1
✈ Chrysler Corporation Scholarship	February 1
✈ David Sarnoff Research Center Scholarship	February 1
✈ Dorothy Lemke Howarth Scholarships	February 1
✈ General Motors Foundation Scholarships	February 1
✈ Ivy Parker Memorial Scholarship	February 1
✈ Judith Resnik Memorial Scholarship	February 1
✈ Lillian Moller Gilbreth Scholarship	February 1
✈ Lockheed Martin Fort Worth Scholarships	February 1

✈ Maswe Memorial Scholarships	February 1
✈ Northrop Corporation Founders Scholarship	February 1
✈ Rockwell Corporation Scholarship	February 1
✈ Stone And Webster Scholarships	February 1
✈ United Technologies Corporation Scholarships	February 1
Graduate Scholarships	
✈ General Motors Foundation Graduate Scholarship	February 1
Reentry Scholarships	
✈ B.K. Krenzer Memorial Reentry Scholarship	May 15
✈ Chrysler Corporation Reentry Scholarship	May 15
✈ Olive Lynn Salembier Scholarship	May 15
Tweet Coleman Aviation Scholarship, The	February
Student Pilot Scholarships	Nofity
Whirly-Girls Scholarships	October 31
Women in Aviation, International	
General Scholarships	
✈ Airbus Leadership Grant	March
✈ Amelia Earhart Society Career Enhancement Scholarship	March
✈ Boeing Company Career Enhancement Scholarship	March
✈ Colleen Barrett Aviation Management Scholarship	March
✈ Flight Training Magazine Scholarships	March
✈ Women in Aviation. International Achievement Awards	March
Flight Scholarships	
✈ Airbus A320 Type Rating Certificate Scholarships	March
✈ American Airlines B-727 Flight Training Award	March
✈ Cessna Aircraft Company Private Pilot Scholarship	March
✈ Jeppesen Sanderson Company Private Pilot Scholarship	March
✈ Northwest Airlines Type Rating Award	March
✈ SimuFlite Citation II Corporate Aircraft Training Scholarship	March
✈ United Airlines Type Rating Scholarship	March
Maintenance Scholarship	
✈ Aircraft Electronics Association Aviation Maintenance Scholarship	March
✈ ATP Maintenance Technician of the Year Award	March
✈ Bombardier Challenger Initial Maintenance Scholarship	March
✈ SimuFlite Maintenance Scholarship	March
Women's Transport Seminar Undergraduate Scholarship	February
Women Military Aviators, Inc. Scholarship	June
Women of the National Agricultural Aviation Association	Nofity
Women Soaring Pilots Association	Nofity

LOANS

Army Aviation Association of America Scholarship Foundation	April
Airmen Memorial Foundation Federal Student Loan Program	Nofity
AOPA's Flight Fund	Nofity
AVCO Financial Services	Nofity
Career Education Loan Program	Nofity
Education Resources Institute	Nofity

Education Credit Corporation	Nofity
Federal Employees Education Association	Nofity
Hattie M. Strong Foundation Student Loan	Nofity
Knights Templar Special Low Interest Loans	Nofity
MBNA - AOPA Gold Option Loan	Nofity
Nellie Mae	Nofity
The Retired Officers Association	Nofity

THIS PAGE INTENTIONALLY LEFT BLANK

Cooperative Education & Internship Programs

What Is A Cooperative Education Program?
Cooperative Education Program (Co-op) is an academic program that gives the student practical work experience prior to graduation. It is a planned and progressive educational program that offers many advantages to all who participate. By combining your academic studies with on-the-job experience, this program helps you get the experience you need to obtain the job of your choice.

You can apply classroom theories to interesting work programs and projects that will give you the real world realities of your career field. Co-op programs also give you the opportunity to explore your options before choosing a career. Eligibility requirements and schedules varies from school to school. More specific information can be obtained from your academic advisor or career center. Companies, organizations, schools, and students design the schedules that works best for them. Co-op jobs usually require a minimum commitment of two separate work periods (semester/quarter) which could be extended throughout college.

What Is An Internship?
An internship is an educational experience in an environment providing field application of what a student learned in class. Internships are arranged for students who are undertaking a study involving both academic and applied experience. They are also arranged for students to receive a certain number of credits for the project.

The number of credits for an internship will be determined by the faculty supervisor, subject to approval by the academic dean. There are several factors that will be considered and the number of hours per week the student will spend working under the supervision of the field supervisor. Internships are usually limited to one work period (typically, summer semester/quarter) with a given employer.

Evaluation of the internship by the faculty supervisor will be made under the same standards that apply to regular course work. The internship will be presented either by reports, papers, presentation, etc. for evaluation. The supervisor will also provide an evaluation of the student's activities and performance.

Reasons For Applying To A Cooperative Education Or Internship Program:
- ✈ You can test your career choice through real world experiences
- ✈ You will have professional experience
- ✈ You can earn money for college expenses
- ✈ You can earn academic credit for your intern or co-op experience
- ✈ You will be more marketable than your competitors who does not have any co-op experience
- ✈ You can gain practical experience in your chosen career field
- ✈ You will have the opportunity to apply skills and knowledge learned in the classroom to actual on-the-job experiences.
- ✈ You may receive college credit toward graduation requirement.
- ✈ You will have work experience to list on resumes and employment applications.
- ✈ And Much More

How To Apply For An Internship Or Co-Operative Education Program
There are several steps to consider when you are seeking an internship or co-operative education position. Each step is important and requires a lot of planning and preparation. Your resume, cover letter, interview, and follow-up letter are very important steps. There is a brief explanation of each step to assist you in your search. There are numerous books on writing effective letters, resumes, and interview preparation tips located at a local library or bookstores.

What Is The Purpose Of A Resume?
The purpose of a resume is: 1. Help sell yourself, 2. Relate your skills and achievements to the job, 3. Focus on the organization, and 4. Get your foot in the door. In summary, you are outlining your work experience, education, and skills on a sheet of paper. Writing an effective resume is the **#1** tool in a job search.

Role of a resume:
- ✈ Its goal is to convince a potential employer to contact you and offer you an interview.
- ✈ It is designed to sell your ability and potential.
- ✈ It should focus on what you can do for the employer, based on evidence of past accomplishments.
- ✈ It should be clear, concise, well organized, easy to read, and one page in length
- ✈ It should include your basic identification, education, work experience, honors and achievement, flight experiences, and certificates / license.
- ✈ Tailors the company and the job position desired or offered (Develop multiple resumes)

A resume that sells should include:
- ✈ Examples of productivity
- ✈ Patterns of accomplishments & upward movement
- ✈ Examples that present you as a team player
- ✈ Evidence of stability & direction

Resume Preparation Tips:
- ✈ An average resume gets about 17 seconds to make its first impression.
- ✈ Information closest to the left and right margins stands out better than that in the middle of the page.
- ✈ The first and last items in a list are those that are noted more quickly than those embedded in the center of the list.
- ✈ Information in columns can be scanned and remembered better than that in paragraphs. Avoid paragraphs.
- ✈ Word phrases are usually better than complete sentences.
- ✈ Use bullets to call attention to important points on your resume.

What Is The Purpose Of A Coverletter?

The purpose of a coverletter is: 1. Introduce yourself to the company, 2. State your objectives, and 3. Sell yourself into an interview. The cover letter follows the basic business letter format. It should be neatly typed on good paper that matches the color and style of your resume.

This is your opportunity to sell yourself to the prospective employer by highlighting your education, experiences, and qualities. Keep in mind the role of the receiver. Remember, the cover letter will serve as a **SALES PRESENTATION** to the reader.

Include the following information within your coverletter:
1. Your address
2. Date
3. Sponsor's name, title, and address
4. The first paragraph (opening) should state your purpose of writing the letter and your interest in the position offered.
5. The second paragraph (boby) highlights those qualifications and interests related to the position you are seeking.
6. The third paragraph (conclusion) indicates your follow-up action and how and when you can be reached for an interview.
7. Thank you

When the cover letter is completed. It should answer the following questions:
- ✈ Tells your prospective employer what you can do for that company .
- ✈ Why you feel you are qualified for the position.
- ✈ Highlights important points in your resume
- ✈ Expand on any past experience that suits the particular internship or co-op for which you are applying.
- ✈ Focuses on what you can contribute to the employer
- ✈ Ask for an interview

Interviewing Process
This section is designed to help you prepare for the interview process. It is important that you are prepared to answer the questions asked during the interview. Having information about the company before the actual interview will help you prepare for any questions regarding the company. This will show the interviewer that you have done your homework (research) on the company.

In an interview, employers do more than check your overall appearance and attitude. They want to hear you speak about your abilities, career objectives, strengths, weaknesses, and potential contributions to their company. After the interviewer has asked questions, you will be expected to ask questions regarding the position you are applying for. Interest is shown by asking relevant and pertinent questions. The interview stage of your search is the most crucial.

Be ready to **SELL YOURSELF**.

Outline For Researching Yourself And The Company

Know yourself and be able to explain them:
- Personality strengths and weaknesses
- Intellectual strength and weaknesses
- Communicative strengths and weaknesses
- Accomplishments and failures
- Professional strengths and weaknesses
- On-the-job strengths and weaknesses
- Reasons for leaving past positions
- What you want in a position
- What are your most valued wants and needs

It is very important to learn as much as possible about the company or organization. Talk to any employees of the company (if you don't know anyone, call the secretary), research the company through the local Chamber of Commerce, professional or trade publications, local newspapers, and the internet.

Know your field
- History
- Developments
- Trends
- Leaders
- Problems
- Education/training requirements

Know the organization
- Products and services (old and new)
- Location of plants, offices, divisions
- Expansion or merger plans
- Reputation
- History
- Major competitors
- Financial status

Know the position
- Responsibilities and duties
- Expectations
- Required or necessary experiences
- Necessary personal characteristics
- Type of supervision
- Advancement potential
- Location
- Starting date
- Job security
- Benefits
- Training
- Salary and/or commissions
- Relocation possibilities and probabilities
- Rate of turnover
- Age of peers and supervisors

Know what happens during the interview process
- Questions the interviewer may ask
- Types of interviews you might encounter

Sample Questions You May Be Asked By The Interviewer
Ask for clarification of questions when necessary. Keep your responses positive, reinforcing your specific interests and qualifications.

- Why do you want to work for our company?
- Why are you applying for this internship or co-op position?
- What special skills/traits do you obtain that qualify you for this position?
- What do you hope to obtain from this co-op or internship?
- Would you rather work alone or in a group? Explain.
- Are you able to work without close supervision? Explain.
- What qualifies you to work here?
- What are your personal/work strengths and weaknesses, and how do you hope to utilize/improve them?
- Why did you choose our company over others?
- What are your career goals? 1 - 4 years? 4 years and beyond?
- What do you expect to gain and learn from your internship / Co-op?
- Tell me about yourself.
- Please describe any aviation/airport experience?

Sample Questions That You Should Asked The Interviewer
Many of your questions may have been asked during the interview or during your research. Make sure you are attentive during the interview.

1. What are the opportunities for personnel growth?
2. How is an intern or co-op student evaluated?
3. Describe the typical assignment?
4. Tell me about your training program?
5. What are the challenging facets of the job?
6. What are the company's plan for future growth?
7. What is the company's record of employment stability?
8. Is the company stable and financially sound?
9. How has the company fared during the past/current recession?
10. What makes your company different from its competitors?
11. What are the company's strengths and weaknesses?
12. What kind of career opportunities are currently available for my degree and skills?
13. Describe the work environment?
14. How can you utilize my skills?
15. What is the overall structure of the department where the position is located?
16. What qualities are you looking for in your new interns and co-ops?

Cooperative Education & Internship Listing

❒ **American Airlines**
Contact: Scott Hansen, Manager Flight Administration
Flight Academy MD823 - PO Box 619617
DFW Airport, TX 75261-9617
Tel. (817) 967-5291
Fax (817) 967-5031
Website: http://www.aa.com

Deadline(s): Summer - February 1, Fall - June 1, Spring - October 1

Duration: One Semester

Location(s): Boston, Chicago, Dallas/Ft Worth, Los Angeles, Miami, New York, San Francisco, Washington DC

Description: American Airlines consider 20-25 interns each semester. A total of 15-20 schools participate in the program. The internship consists of spending one semester in Dallas, TX at the American Airlines Flight Academy or in a Flight Office at one of the major hubs. Interns support the following departments: Flight Administration, Flight Operations, Technical, International Operations, Crew Communications, Pilot recruitment, Program Development, and American Eagle Operations. Students who successfully complete the internship will be assisted in getting an interview when hiring requirements are met.

Eligibility: Senior or junior status, Minimum 3.0 cumulative GPA, Commercial Certificate! Instrument Rating, Second Class FAA medical with ability to obtain First Class, U.S. citizen or alien with legal right to accept employment in U.S.

School participation: University of North Dakota, Western Michigan University, Utah State University, University of Oklahoma, Louisiana Tech University, Metropolitan State College of Denver, Ohio University, Purdue University, Southeastern Oklahoma State University, Southern Illinois University, Kent state University, Florida Institute of Technology, and Embry-Riddle Aeronautical University (Daytona and Prescott)

Compensation: Unpaid, but will receive college credit.

❒ **Arizona/NASA Space Grant Program**
Contact: Susan A. Brew, Sr. Coordinator
Lunar & Planetary Laboratory
The University of Arizona
Tucson, AZ 85721
Tel. (520) 621-8556
Fax (520) 621-4933
Website: http://seds.org/spacegrant.htm
E-mail: sbrew@seds.org

Deadline: Varies for each University

Duration: One academic year (Awards are renewable at Arizona State University)

Description: Students are assigned to work with faculty, researchers, scientists and engineers in a wide range of space and related disciplines including astronomy, physics, chemistry, geosciences, aerospace and related engineering, planetary sciences, global change-related fields, space biology, science education and science journalism.

Eligibility: Varies for each university. All university programs require U.S. Citizenship. The University of Arizona accepts sophomores and above, the other universities also accept some freshmen. All seek bright and highly motivated students seeking a challenging, hands-on, educational experience.

Compensation: Approximately $6.00/hr

❐ **The Boeing Company**
Laura Sycamore, College Recruiting
PO Box 3707, MC 6H-PR
Seattle, WA 98124
Tel. (425) 237-4599
Fax (425) 234-2568
Website: www.boeing.com
E-mail: submit.resume@boeing.com or laura.k.sycamore@boeing.com

Deadline: Rolling

Duration: 3 months or 6 months

Location(s): Alabama, Arizona, Florida, Kansas, Missouri, Puget Sound, Southern California, Texas

Description: Challenging full-time paid assignments are available. Relocation may be available if applicable. Summer internship candidates are normally reviewed January through March and internship offers extended January through March. Positions varies from region and by division. a student may be apart of the Commercial Airplane Group, Shared Services Group, Space & Communications Group, Military Aircraft, and Missile Systems.

Eligibility: College students in Engineering, Computing, and Business are eligible. A minimum 2.5 GPA is preferred for Engineering and Computing and a minimum 3.3 GPA is preferred for Business. A student must have authorization to work full-time in the United States for other than training purposes.

Compensation: Based on three categories (Engineering, Computing, and Business) and the number of quarters/semesters remaining.

❐ **Cheyenne Airport**
Contact: Martin P. Lenss, Assistant Airport Manager
PO Box 2210
200 East 8th Ave
Cheyenne, Wyoming 82003 - 2210
Tel. (307) 634-7071
Fax: (307) 632-1206

Deadline: April

Description: Intern's tasks will be varied providing exposure to Airport Administration and Finance, Operations, Maintenance, Public Relations, and a wide variety of other duties and responsibilities as assigned. One, three-month Airport Management Intern position. The Starting and ending dates are flexible depending on the individual's college/university schedule. Airport administration may be able to offer rental housing.

Eligibility: Intern position open to candidates currently enrolled in Aviation/Airport Management at the college/university undergraduate level. Computer literacy a plus. Requires strong written and oral communications skills and ability to work independently. The candidate must have at least one semester of undergraduate studies after completing the internship.

Compensation: $6.00/Hr., full-time, No benefits are included.

❐ **Colorado Aviation Management Internship Program**
Department of Transportation
Division of Aeronautics
56 Inverness Dr. East
Englewood, CO 80112-5114
Tel. (303) 792-2160

Description: This program is a state sponsored internship opportunity for undergraduate and graduate students enrolled in aviation-related career areas to have a first hand experience in the field of aviation management. The program 9 month - 1 year internships at the state wide large general aviation and small commercial service airports. State/airport joint funding at a minimums of $10 hour for a 40 - hour week. Interns train under the guidance of airport staff while utilizing the AAAE internship syllabus.

Note: The State of Colorado is not directly involved in the selection process and that these interns are in no way considered employees of the State of Colorado.

❒ Daimler-Benz Aerospace
Posttach 80 1109
81 663 Munich
Tel: +49 89 607 345 15
Fax: +49 89 607 346 67
Website: http:www.DASA.DE

Contact: Pablo Salame Fisher, Mgr. of Human Resource - Marketing

Deadline: Contact Sponsor

Description: 3 to 6 months internships at any of the 36 DASA sites, working on mostly technology and engineering issues. All in universities in Europe.

Eligibility: Student of engineering or business administrative, information technologies or similar. Good communication of German and the English language.

Compensation: Contact Sponsor

❒ EAA Aviation Foundation, Inc.
Education Office
PO Box 3065
Oshkosh, WI 54903-3065
Tel. (414) 426 - 6815

The EA. offers various internship opportunities to students each year. Provided below is a brief description for some of the internships. Contact the Education Office for duration, compensation, and eligibility requirements.

Sandberg Summer Internship - offers an aspiring aviation technician experience with the unique, flying aircraft hangared at EAA's Kermit Weeks Flight Research Center in Oshkosh, Wisconsin.

Doolittle Raiders Internship - enables an educator to work with the EAA's Education Office in developing, organizing and delivering the many youth and education programs organized and presented by the EAA.

The Timken Aviation Studies Internship - is funded by The Louise H. Timken "Young Women In Aviation" Endowment. It provides a college-level practice for a future aviation professional at the EAA Aviation Center in Oshkosh, Wisconsin.

❒ Evansville-Vanderburgh Airport Authority District
Evansville Regional Airport
7801 Bussing Drive
Evansville, IN 47711-6799
Tel. (812) 421-4401
Fax: (812) 421-4412

Contact: Jeff Mulder, A.A.E., Assistant Manager
E-mail: atjam@juno.com

Deadline: April 1, 1998

Duration: June 1, 1998 through August 31, 1998

Description: The Evansville Regional Airport will offer several internship positions this summer for students currently enrolled or recently graduated from an Aviation Program. The interns will be working primarily with the airfield and building maintenance departments, but will also spend some time in the safety building and administration office.

❐ **Illinois Aviation Trades Association Internship Programs**
Contact: Michael A. Pryor
C/O Priester Aviation
Palwaukee Municipal Airport
Wheeling, IL 60090
Tel. (847) 537-1200

Deadline: March

Description: The member companies of the IATA offers wide variety of positions in, but not limited to : Aircraft maintenance, Flight, management, Ramp Service, and Administration. IATA members are located throughout the state of Illinois, including: Champaign, Bloomington, Springfield, Peoria, Galesburg, Chicago, and Danville.

Eligibility: Potential interns must be motivated, self starting individuals seeking to maximize the benefits of participation in such a program. Selection will be based on the individual's academic record, recommendations, and geographic availability of a sponsoring IATA member.

Pay Rate: Contact Sponsor

❐ **Landrum & Brown, Inc.**
Contact: Christine Gerencheer
1021 W. Adams, Suite 200
Chicago, IL 60607
Tel. (312) 421-0500
Fax: (312) 421-6171

Deadline: March

Description: Landrum & Brown, Inc. is a prominent Aviation Consulting firm based in Cincinnati, Ohio with offices in Chicago and Los Angeles. The Environmental Group is one of four practices in the firm that include the facilities and Operations Planning Practice, Terminal Planning Practices, and Financial Planning and Program Implementation Practice.

The Environmental Practice internship will provide students with experience in the development of environmental assessments, environmental impact studies, part 150 studies, and analyses to assist airports with all aspects of environmental planning and compliance.

Eligibility: Candidates should be enrolled in a bachelor level program (Master's program candidates preferred) with a specialization in aviation, geography, environmental sciences, or planning

Pay Rate: Contact Sponsor

❐ **NASA**
NASA Headquarters Higher Education Branch
Mail Code FEH
Washington, DC 20546
Tel. (202) 358-0000

Deadline: Varies, from December 31 for some summer undergraduate programs at the field offices to April 1 for some fellowships.

Description: There are at least 200 different NASA programs, ranging from high school summer jobs to undergraduate and graduate internships to graduate fellowships. The multitude of programs allows students to work at NASA's nine facilities (listed below), and within its dozens of university research labs. NASA has much to offer students who are interested in airplane design and outer space technology. Areas of research include robotics, earth sciences (biology, geology, environmental science, etc.), aerodynamics, biomedicine and biotechnology, materials processing, space propulsion, space structures, and satellite communications.

Eligibility: NASA has an internship for virtually everyone, from high school students at least 16 years of age to college undergraduates to graduate students (including those studying medicine, business, or law) to faculty members. There are also specific programs for minority undergraduates and graduate students in addition to minority programs at the field centers. Some programs require a minimum 3.0 GPA. International applicants eligible.

Compensation: Varies considerably. Pay depends on major, class level, and experience. Some interns are also provided with travel stipends and free housing. In the case of Goddard's Summer Institute, a few government cars are also at interns' disposal.

Note: To receive information on programs specific to particular NASA centers, including part-time and summer work for high school students, as well as the NASA Cooperative Education Program for high school students and college undergraduates, contact the offices at the addresses and phone numbers listed below.

NASA Ames Research Center University Affairs Office
Code 241-3
Moffett Field, CA 94035
Tel. (415) 604-5802

NASA Goddard Space Flight Center University Programs
Mail Stop 160
Greenbelt Road
Greenbelt, MD 20771
Tel. (301) 286-9690

Jet Propulsion Laboratory
183-900 Educational Affairs Office
4800 Oak Grove Drive
Pasadena, CA 91109-8099
Tel. (818) 354-8251

NASA Johnson Space Center University Programs
Mail Stop AHU
Houston, TX 77058
Tel. (713) 483-4724

NASA Kennedy Space Center University Program Manager
Mail Code HM-CIC
KSC, FL 32899
Tel. (407) 867-7952

NASA Langley Research Center Office of Education
Mail Stop 400
Hampton, VA 23681-0001
Tel. (804) 864-4000

NASA Lewis Research Center Educational Programs
Mail Stop 7-4 21000
Brookpark Road
Cleveland, OH 44135
Tel. (216) 433-2957

NASA Marshall Space Flight Center University Affairs
Mail Stop DS01
MSFC, AL 35812

Tel. (205) 544-0997

NASA Stennis Space Center Management Operations
University Affairs Office
Mail Code MA00
SSC, MS 39529
Tel. (601) 688-3830

❒ NATIONAL AIR AND SPACE MUSEUM INTERNSHIPS
National Air and Space Museum
Educational Services Division
Attn: Student Services Coordinator
Room P700, MRC 305
Washington, DC 20560
Tel. (202) 786-2106
TTY: (202) 357-1505

Deadline: February for the summer internship; June for the fall internship; September for the spring internship.

Purpose: To provide internship opportunities at the Smithsonian Institution's National Air and Space Museum. Internships available in areas such as aviation, astronomy, geology, space science, etc.

Eligibility: This program is open to undergraduate and graduate students who have been studying museology, history, art, aviation, space science, photography, journalism, or education.

Award(s) & Amount(s): A stipend is paid, unless the intern receives academic credit.

Duration: At least 10 weeks.

❒ National Air Transportation Association
Contact: Douglas Carr
Specialist, Government and Industry Affairs
4226 King Street
Alexandria, VA 22302
Tel. (800) 808 - 6282

Deadlines: November for Spring Internship; March for Summer & fall internships

Description: NATA is the public policy group representing thousands of aviation businesses including fixed based operators (FBOs), air charter and commuter operators, flight schools, maintenance repair stations, and airline-service companies before Congress and the federal agencies. Selected applicants will participate in regulatory and legislative activities before the highest levels of government at the Association's national headquarters, just outside Washington, D.C.

Eligibility: Enrolled in a college/university aviation program with a junior or higher standing, at least one semester of college work remaining following the internship with NATA, Prior work experience in FBO, air charter, or airport operations desirable, Strong written and verbal communication skills, and aviation-related extracurricular activity involvement.

Pay Rate: One time $1,100 stipend. All other expenses are the responsibility of the selected applicant.

❒ National Space Society
Contact: Pat Dasch, Executive Director
600 Pennsylvania Ave., S.E.
Suite 201
Washington, DC 20003
Tel. (202) 543-1900
Fax (202) 546-4189

Website: http://www.nss.org
E-mail: nsshq@nss.org

Deadline: March 15

Duration: 6 week minimum

Description: Interns participate in a wide variety of activities ranging from general office duties to helping with publication of the society's magazine AD Astra, to assisting with various special events on Capitol Hill, etc. College students from all backgrounds are accepted.

Compensation: Unpaid

❒ Snohomish County Airport
3220 100th Street SW
Everett, WA 98204 - 1390
Tel. (206) 353-2110

Contact: Bruce Goetz

Deadline: April

Description: The internship begins with a 6 month internship and can include up to 18 additional months on the airport staff. A follow-on regular position with Snohomish County Airport is not guaranteed. Areas covered in the internship program may include but not limited to: Administration, Finance, Land Development, Operations, Maintenance, Engineering, Facilities and Environmental programs.

Eligibility: Requires a Bachelor's degree in aviation management/administration or aviation related field. Female and minority candidates are encouraged to apply. Students in final quarter/semester may apply. Advanced knowledge of PC's and computer networks preferred.

Pay Rate: $11.00/Hr., Medical benefits may be available after the 6 month internship.

❒ Wayne County Detroit Metropolitan Airport
Airport Summer Internships
Airport Director's Office - Mezzanine
L.C. Smith Terminal
Detroit, MI 48242
Tel: (313) 942-3559
Fax: (313) 942-3793

Contact: Renee M. Filer

Deadline: April

Description: Reporting on passengers and operation statics, airport and faculty surveys, responding to general and public information requests, preparing feasibility studies, and assisting in the coordination of Airport - related media events.

Six positions are available for each internship period beginning in May, September and January. Interns are assigned to the areas of airport Administration, Operations, Facilities Planning, Community Relations, and Training Departments. The average period of employment if 4 months.

Eligibility: Students working towards a B.S. in an accredited Aviation Program at the Junior level or higher as well as recent graduates of Aviation programs are encouraged to apply.

Pay Rate: $7.00/Hr.; Bi-weekly; Work Hours: Monaday - Friday 8:00am - 4:30 pm

FAA Aviation Education Representatives

Department of Transportation/FAA
Phillips S. Woodruff, AHT-100
Director of Aviation Education
800 Independence Ave., S.W.
Washington, DC 20591
Tel. (202) 267-3788

Headquarters

Aviation Education Division
Terry White
Patsy Vicks
Latisha Ferguson
Office of Traiinig & Higher Education
Aviation Education Divison
400 7th Street, SW, Rm Pl-100
Washington, DC 20590
Tel. (202) 366-7500
Fax (202) 366-3786

Aeronautical Center
Robert Hoppers, AMC-5
Rm 356, Headquarters Bldg.
PO Box 25082
Oklahoma City, OK 73125
Tel. (405) 954-7500
Fax (405) 954-4551

Technical Center
Carleen Genna, ACM-120
Atlantic City International Airport
Human Resource Management Division
Atlantic City, NJ 08405
Tel. (609) 485-6626
Fax (609) 485-4391

Center for Management Development
Larry Hedman, CMD-373
4500 Palm Coast Parkway, SE
Palm Coast, FL 32137
Tel. (904) 446-7126
Fax (904) 446-7201

Alaskan Region
Mary Lou Dordan, AAL-5B
222 West 7th Ave., Box 14
Anchorage, AK 99513-7587
Tel. (907) 271-5293
Fax (907) 276-7261
State: Alaska

Central Region
Maria Navarro, ACE-5
601 East 12th St.
Federal Building, Rm 1501
Kansas City, MO 64106
Tel. (816) 426-5836
Fax (816) 426-5434
States: Iowa, Kansas, Missouri, Nebraska

Eastern Region
Jim Szakmary, AEA-17
JFK International Airport
Federal Building #111
Jamaica, NY 11430
Tel. (718) 553-1056
Fax (718) 553-0058
States: Delaware, District of Columbia, Maryland, New Jersey, New York, Pennsylvania, Virginia, West Virginia

Great Lakes Regions
Lee Carlson, AFL-14B
O'Hara lake Office Center
2300 East Devon Ave
Des Plaines, IL 60018
Tel. (708) 294-7042
Fax (708) 294-7642 or 7691
States: Illinois, Indiana, Michigan, Minnesota, North Dakota, Ohio, South Dakota, Wisconsin

New England
Shelia Bauer, ANE-45
12 New England Executive Park
Burlington, MA 01803
Tel. (617) 238-7378
Fax (617) 238-7380
States: Connecticut, Maine, New Hampshire, Rhode Island, Vermont, Massachusetts

Northwest Mountain Region
Herman Payton, ANM-14A
1601 Lind Ave, SW
Renton, WA 98055
Tel. (206) 227-2079
Fax (206) 227-1010
States: Colorado, Idaho, Montana, Oregon, Utah, Washington, Wyoming

Southern Region
Opal Neely
1701 Columbia Ave
College Park, GA 30337
Tel. (404) 305-5386
Fax (404) 305-5312
States: Alabama, Florida, Georgia, Kentucky, Mississippi, North Carolina, South Carolina, Tennessee, Puerto Rico, Virgin Islands

Southwest Region
Debre Myers, ASW-1SB
Aviation Education Program
Manager
Federal Aviation Administration
Ft. Worth, TX 76193
Tel. (817) 222-5833
Fax (817) 222-5950
States: Arkansas, Louisiana, New Mexico, Oklahoma, Texas

Western-Pacific Region
Hank Verbais
PO Box 92007
Worldway Postal Center
Los Angeles, CA 90009
Tel. (310) 297-0556
Fax (310) 297-0706
States: Arizona, California, Nevada, Hawaii

State Aviation & Airport Departments

This section provides you with a listing of state aviation and airport departments. These departments may offer airport-related scholarships, grants, internships, coops, or loans to students. These are usually in the aviation safety/education bureaus or in the airport engineering/planning bureau. Some of the internships and coops may or may not compensation. These state departments may be restricted to their legal residents. However, they may also be available to out-of-state students who may be or are attending public or private colleges or universities within the state. These departments are aware of many opportunities for career advancements.

Contact: Mr. John Eagerton, Director
Alabama Department of Aeronautics
770 Washington ave., Suite 544
Montgomery, AL 36130
Tel. (334) 242-4480
Fax (334) 240-3274

Contact: Mr. Roger Maggard
Alaska Statewide Aviation
P.O. Box 196900
Anchorage, AK 196900
Tel. (907) 266-1650
Fax (907) 243-1512

Contact: Ms. Doris Mecham
Airport Development Manager
Arizona Division of Aeronautics
P.O. Box 13588, MD426M
Phoenix, AZ 85002-3588
Tel. (602) 255-7691
Fax (602) 407-3007

Contact: Mr. Fred Dodd
Program Administrator
Arkansas Dept. of Aeronautics
1 Airport Drive
Little Rock, AK 72202
Tel. (501) 376-6781
Fax (501) 378-0820

Contact: Ms. Teresa Ishikawa
Airport Inspector
California Aeronautics Program
MS-40
P.O. Box 942874
Sacramento, CA 94273-0001
Tel. (916) 322-9942
Fax (916) 327-9093

Contact: Ms. Caroline Scott
Aviation Education
Colorado Division of Aeronautics
56 Iverness Dr. East
Englewood, CO 80112-5114
Tel. (303) 792-2158
Fax (303) 792-2180

Contact: Ms. Kathy Maznicki
Grants Administrator

Connecticut Dept of Transportation
24 Wolcott Hill Rd.
Wethersfield, CT 06129-0801
Tel. (203) 292-2014
Fax (203) 292-2090

Contact: Mr. Harry VanDenHeuvel
Delaware Office of Aeronautics
P.O. Box 778
Dover, DE 19903
Tel. (302) 739-5712 or 3264
Fax (302) 739-5711

Contact: Mr. Bill Sherry, Manager
Sr. Transportation Planner
Florida Aviation Office
605 Suwannee St., MS-46
Tallahassee, FL 32399-0450
Tel. (904) 488-8444
Fax (904) 922-4942

Contact: Mr. Ed Ratigan
Georgia Aviation Programs
Georgia Department Aviation
276 Memorial Drive SW
Atlanta, GA 30303-3743
Tel. (404) 651-5208 or 9200
Fax (404) 651-5209

Guam Airport Division
P.o.Box 8770
Tamuning, Guam 96931
Tel (011) 671-646-0300

Contact: Mr. Stanford Miyamoto
Aviation Planning
Hawaii Airports Division
Honolulu International Airport
400 Rogers Blvd., Ste 700
Honolulu, HI 96819-1880
Tel. (808) 838-8701 or 8600
Fax (808) 838-8750

Contact: Mr. Tim Peterson
General Aviation Officer
Idaho Deparment of Transportation
P.O. Box 7129
Boise, ID 83707-1129

Tel. (208) 334-8780 or 8788
Fax (208) 334-8789

Contact: Mr. Jim Bildilli
Illinois Division of Aeronautics
Capital Airport
One Longhorn Bond Dr.
Springfield, IL 62707-8415
Tel. (217) 785-8516 or 8515
Fax (217) 785-4533

Contact: Mr. Troy Allen
Chief, Bureau AvEd & Safety
Indiana Aeronautics Section
100 N. Senate Ave., Rm N901
Indianapolis, IN 46204-2217
Tel. (317) 232-1494 or 1477
Fax (317) 232-1499

Contact: Ms. Kathie Robinson
Project Manager
Iowa Plan/Program Coordination
Iowa Plan/Program Divison
100 E. Euclid Ave., Ste 7
Des Moines, IA 50313-4564
Tel. (515) 237-3314 or 3311
Fax (515) 237-3323

Contact: Mr. David Cushing
Education Publications
Kansas Division of Aviation
Docking State Office Bldg
Room 726 North
Topeka, KS 66612-1568
Tel. (913) 296-7498 or 7449
Fax (9130 296-3833

Contact: Mr. Bob Bodner
Policy & Development Planner
Kentucky Office of Aeronautics
Kentucky Divison of Divison
125 Holmes St.
Frankfort, KY 40622
Tel. (502) 564-4480
Fax (502) 564-7953

Contact: Mr. Anthony Culp
Executive Director
Louisiana Aviation Division
P.O. Box 94245
Baton Rouge, LA 70804-9245
Tel. (504) 379-1242
Fax (504) 379-1961

Contact: Mr. Ron Roy, Director
Aviation Director
Maine Air Transportation Division

Maine Office of Passenger Transportation
State House Station #16
Augusta, ME 04333
Tel. (207) 287-3186
Fax (207) 287-2805

Contact: Mr. Bruce Mundie
Maryland Aviation Administration
P.O. Box 8766
Baltimore/BWI, MD 21240
Tel. (410) 859-7064or 7060
Fax (410) 850-4729

Contact: Ms. Linda Atteratta
Director Reg. Aviation Assistance
Massachusetts Aeronautics Commission
Boston, MA 02116-3966
Tel. (617) 973-8893or 8881
Fax (617) 973-8889

Contact: Mr. Tom Krashen
Aviation Planner
Michigan Bureau of Aeronautics
2700 E. Airport Service Dr.
Lansing, MI 48906-2171
Tel. (517) 335-8396or 9943
Fax (517) 321-6422

Contact: Mr. Gordon Hoff
Aviation Education
Minnesota Aeronautics Office
222 E. Plato Blvd.
St. Paul, MN 55107-1618
Tel. (612) 297-7652or 296-8046
Fax (612) 297-5643

Contact: Mr. Wayne Caldwell
Manager, General Aviation
Mississippi Aeronautics Division
P.O. Box 1850
Jackson, MS 39215-1850
Tel. (601) 359-7850
Fax (601) 359-7855

Contact: Ms. Alice Graham
Aeronautics Engineer
Missouri Aviation Section
P.O. Box 270
Jefferson City, MO 65102
Tel. (314) 751-2589or 526-7912
Fax (314) 526-4709

Contact: Mr. Jeannie Lesnik
Aviation Planner
Montana Aeronautics Division
P.O. Box 5178
Helena, MT 59604

Tel. (406) 444-2506
Fax (406) 444-2519

Contact: Mr. Neil Vernon
Chief, Safety & Education Bureau
Nebraska Department of Aeronautics
Lincoln, NE 68501
Tel. (402) 471-2371
Fax (402) 471-2906

Contact: Mr. Dennis Taylor
Aviation Specialist
Nevada Department of Transportation
1263 S. Stewart St.
Carson City, NV 89712
Tel. (702) 687-5653
Fax (702) 687-4846

Contact: Mr. Doug Teel
Aviation Planner
New Hampshire Division of Aeronautics
Municipal Airport
65 Airport Rd.
Concord, NH 03301-5298
Tel. (603) 271-2551or 1676
Fax (603) 271-1689

Contact: Mr. James Varanyak
Aviation Safety & Education Specialist
New Jersey Division of Aeronautics
1035 Parkway Ave., CN-610
Trenton, NJ 08625
Tel. (609) 530-2914or 2080
Fax (609) 530-5719

Contact: Mr. Wayne York
Aviation Education Contact
New Mexico Aviation Division
P.O. Box 1149
Santa Fe, NM 87504-1149
Tel. (505) 827-1525
Fax (505) 827-1531

Contact: Ms. Laura Lemire
Flying Safety Coordinator
New York Aviation Division
Nnew York Aviation Servics Burea
1220 Washington Ave.
Albany, NY 12232
Tel. (518) 457-2822
Fax (518) 457-9779

Contact: Mr. Mike Wright
Airport Development Specialist
North Carolina Division of Aviation
P.O. Box 25201
Raleigh, NC 27611
Tel. (919) 571-4904
Fax (9190 571-4908

Contact: Mr. Roger Pfeiffer
North Dakota Aeronautics Commission
Bismark, ND 58502
Tel. (701) 328-9652or 9650
Fax (701) 328-9656

Contact: Mr. Eric Smith
Assistant Director
Ohio Division of Aviation
Ohio Office of Aviation
2829 W. Dublin-Granville Rd.
Columbus, OH 43235
Tel. (614) 793-5048or 5042
Fax (614) 793-8972

Contact: Ms. Pamela Goodyear
Aviation Specialist-Publications
Oklahoma Aeronautics Commission
200 NE 21st St., Rm B7, 1st Fl
Oklahoma City, OK 73105
Tel. (405) 521-2377
Fax (405) 521-2379

Contact: Mr. Gerald Eames
Aviation Education
Oregon Aeronautics Section
3040 25th Street SE
Salem, OR 97310-0100
Tel. (503) 378-4887or 4882
Fax (503) 373-1688

Contact: Mr. Randy Hicks
Airport Standards & Certifications
Pennsylvania Dept. of Transportation & Aviation
Pennsylvania Bureu of Aviation
Forum Place, 8th Floor, 555 Walnut
Harrisburg, Pa 17101-1900
Tel. (717) 948-3915or 705-1200
Fax (717) 948-3527

Puerto Rica Ports Authority
Po Box 362829
San Jaun, 00936-2829
Tel (787)729-8804

Contact: Mr. Eugene Tansey
Manager, Resource Management
Rhode Island Airport Corporation
2000 Post Rd.
Warwick, RI 02886-1533
Tel. (401) 737-4000
Fax (401) 732-4953

Contact: Mr. William Carlisle, Director
Deputy Executive Director
South Carolina Division of Aeronautics
P.O. Box 280068
Columbia, SC 29228
Tel. (803) 822-5400
Fax (803) 822-4312

Contact: Mr. David Jagim
South Dakota Aviation Office
700 E. Broadway Ave.
Pierre, SD 57501-2586
Tel. (605) 773-5037
Fax (605) 773-3921

Contact: Ms. Kathy Sloan
Program Manager
Tennessee Aeronautics Divsion
P.O. Box 17326
Nashville, TN 37217
Tel. (615) 741-3208
Fax (615) 731-4959

Contact: Mr. William Gunn
Aviation Education Contact
Texas Division of Aviation
125 E. 11th Street
Austin, TX 78701-2483
Tel. (512) 476-9262or 416-4501
Fax (512) 479-0294

Contact: Mr. Steve Walrath
Manager, AVED Programs
Utah Aeronautical Operations Division
135 North 2400 West
Salt Lake City, UT 84116
Tel. (801) 533-5057
Fax (801) 533-6048

AVED & Safety Specialist
Contact: Ms. Kathy Shambo
Vermont Division of Rail, Air & Public Transportation
133 State Street
Montpelier, VT 05633
Tel. (802) 828-2087or 2093
Fax (802) 828-2829

Contact: Ms. Betty Wilson
Aviation Education
Virginia Department of Aviation
5702 Gulfstream Rd.
Richmond Int'l Airport 23250-2422
Tel. (804) 236-3625
Fax (804) 236-3635

Contact: Mr. Newell Lee
Aviation Education
Washington Aviation Division
8900 E. Marginal Way S.
Seattle, WA 98108
Tel. (206) 764-4131
Fax (206) 764-4001

Contact: Mr. Randall Biller
Aeronautics Program Specialist
West Virginia Department of Transportation
West Virginia Aeronautics Commission
State Capitol Complex
Bldg. 5, Rm A-931
Charleston, WV 25305
Tel. (304) 558-0330
Fax (304) 558-0333

Contact: Mr. Tomas Thomas
Acting Aeronautics Director
Wisconsin Bureau of Aeronautics
P.O. Box 7914
Madison, WI 53707-7914
Tel. (608) 266-2023or 2480
Fax (608) 267-6748

Contact: Mr. Ed Pew, Senior Pilot
Aviation Management & Education
Wyoming Aeronautics Division
P.O. Box 1708
Cheyenne, WY 82003-1708
Tel. (307) 777-4880
Fax (307) 637-7352

THIS PAGE INTENTIONALLY LEFT BLANK

Space Grant Consortium Directors

This section provides you with a listing of Space Grant Consortium Directors and Departments. The goal of the Space Grant Consortium programs is to encourage interdisciplinary training and research, to train professionals for careers in aerospace science, technology, and allied fields, and to encourage individuals from underrepresented groups to consider careers in aerospace fields.

Space related fields include any academic discipline or field of study (including the physical, natural, and biological sciences; engineering; education; economics; business; sociology; behavioral sciences; computer science; communications; law; international affairs; and public administration) that is concerned with or that is likely to improve the understanding, assessment, development, and utilization of space.

Individuals from underrepresented groups, specifically African-Americans, Hispanics, American Indians Pacific Islanders, Asian Americans, and women of all races who have interest in the aerospace fields, are encouraged to apply.

These departments may be restricted to their state's legal residents. However, they may also be available to out-of-state students who may be or are attending public or private colleges or universities within the state. These departments are aware of many opportunities for career advancements such as scholarships, grants, internships, cooperative education programs, and loans. Contact a director or the Consortium office for more information and deadline dates.

Directors & Departments

Alabama Space Grant Consortium

Dr. John C. Gregory
Director, Alabama Space Grant Consortium
Chemistry Department
College of Science
Materials Science Building, Room 205
Huntsville , AL 35899
Tel: (205) 890-6028
Fax: (205) 890-6061
Email: jcgregory@matsci.uah.edu

Alaska Space Grant Consortium

Dr. Joseph G. Hawkins
Director, Alaska Space Grant Program
University of Alaska, Fairbanks
Electrical Engineering Department
College of Science, Engineering, and Mathematics
223 Duckering Bldg, PO. Box 755919
Fairbanks , AK 99775-5919
Tel: (907) 474-5206
Fax: (907) 474-5135
Email: ffjgh@uaf.edu

Arizona Space Grant Consortium

Dr. Eugene H. Levy
Director, Arizona Space Grant Consortium
University of Arizona
Lunar & Planetary Laboratory, Planetary Sciences
Gould-Simpson, Room 1025
Tucson , AZ 85721
Tel: (520) 621-4090
Fax: (520) 621-8389
Email: ehl@u.arizona.edu

Arkansas Space Grant Consortium

Dr. M. Keith Hudson
Director, Arkansas Space Grant Consortium
University of Arkansas, Little Rock
College of Science and Eng. Techn.
ETAS 125
2801 South University Avenue
Little Rock , AR 722041099
Tel: (501) 569-8212
Fax: (501) 569-8039
Email: mkhudson@ualr.edu

California Space Grant Consortium

Dr. Michael Wiskerchen
Director, California Space Grant Consortium
University of California, San Diego
California Space Institute
9500 Gilman Drive, 0524
La Jolla , CA 92093
Tel: (619) 534-5869
Fax: (619) 5347840
Email: mwiskerchen@ucsd.edu

Colorado Space Grant Consortium

Ms. Elaine R. Hansen
Director, Colorado Space Grant Consortium
University of Colorado, Boulder
Space Grant College
Engineering & Applied Science Department
Engineering Center, Room 1B-76
Campus Box 520
Boulder , CO 80309-0520
Tel: (303) 492-3141
Fax: (303) 492-5456
Email: elaine@rodin.colorado.edu

Connecticut Space Grant Consortium

Dr. Ladimer S. Nagurney
Director, Connecticut Space Grant Consortium
University of Hartford
Department of Electrical Engineering
United Technology Hall
West Hartford , CT 06117
Tel: (860) 768-4866
Fax: (860) 768-5073
Email: nagurney@uhavax.hartford.edu

Delaware Space Grant Consortium

Dr. Norman F. Ness
Director, Delaware Space Grant Consortium
Bartol Research Institute
Bartol Research Institute, University of Delaware
217 Sharp Laboratory
Newark , DE 19716-4793
Tel: (302) 831-8116
Fax: (302) 831-1843
Email: nfness@bartol.udel.edu

District of Columbia Space Grant Consortium

Dr. John Logsdon
Director, D.C. Space Grant Consortium
George Washington University
Space Policy Institute
2013 G Street, NW
Stuart Hall, Suite 201
Washington , DC 20052
Tel: (202) 994-2615
Fax: (202) 994-1639
Email: logsdon@gwis2.circ.gwu.edu

Florida Space Grant Consortium

Dr. Humberto Campins
Director, Florida Space Grant Consortium
University of Florida
Department of Astronomy
222 Space Sciences Research Building
Gainesville , FL 32611-2055
Tel: (352) 392-6750
Fax: (352) 392-3456
Email: fsgc@astro.ufl.edu

Georgia Space Grant Consortium

Dr. Erian Armanios
Director, Georgia Space Grant Consortium
Georgia Tech
Department of Aerospace Engineering
Atlanta , GA 30332-0150
Tel: (404) 894-8202
Fax: (404) 894-9313
Email: erian.armanios@aerospace.gatech.edu

Hawaii Space Grant Consortium

Dr. G. Jeffrey Taylor
Director, Hawaii Space Grant Consortium
University of Hawaii, Manoa Campus, Oahu
Department of Planetary Geosciences
School of Ocean and Earth Science and Technology
2525 Correa Road
Honolulu , HI 96822
Tel: (808) 956-3899
Fax: (808) 956-6322
Email: gjtaylor@pgd.hawaii.edu

Idaho Space Grant Consortium

Dr. Rick Gill
Director, Idaho Space Grant Consortium
Department of Mechanical Engineering
University of Idaho
Moscow , ID 83844-1011
Tel: (208) 885-7018
Fax: (208) 885-6645
Email: rgill@uidaho.edu

Illinois Space Grant Consortium

Dr. Wayne C. Solomon
Director, Illinois Space Grant Consortium
University of Illinois/Urbana-Champaign
Department of Aero/Astronautic Engineering
306 Talbot Laboratory
104 South Wright Street
Urbana , IL 61801-2935
Tel: (217) 244-7646
Fax: (217) 244-0720
Email: wsolomon@uiuc.edu

Indiana Space Grant Consortium

Dr. Dominick Andrisani, II
Director, Indiana Space Grant Consortium
Purdue University
School of Aeronautics and Astronautics
1282 Grissom Hall
West Lafayette , IN 47907-1282
Tel: (765) 494-5135
Fax: (765) 494-0307
Email: andrisan@ecn.purdue.edu

Iowa Space Grant Consortium

Mr. William J. Byrd
Director, Iowa Space Grant Consortium
Dept. of Aerospace Eng./Engineering Mechanics
408 Town Engineering Building
Ames , IA 50011-3231
Tel: (515) 294-3106
Fax: (515) 294-3262
Email: wbyrd@iastate.edu

Kansas Space Grant Consortium

Dr. David R. Downing
Director, Kansas Space Grant Consortium
University of Kansas
Department of Aerospace Engineering
2004 Learned Hall
Lawrence , KS 66045
Tel: (913) 864-4265
Fax: (913) 864-3597
Email: ksgc@aerospace.ae.ukans.edu

Kentucky Space Grant Consortium

Dr. Richard Hackney
Director, Kentucky Space Grant/NASA EPSCoR
Western Kentucky University
Department of Physics & Astronomy
One Big Red Way
Bowling Green , KY 42101-3576
Tel: (502) 745-4156
Fax: (502) 745-6471
Email: ksgc@wkuvx1.wku.edu

Louisiana Space Grant Consortium

Dr. John Wefel
Director, Louisiana Space Grant Consortium
Louisiana State University
Department of Physics and Astronomy
277 Nicholson Hall
Baton Rouge , LA 70803-4001
Tel: (504) 388-8697
Fax: (504) 388-1222

Email: wefel@phepds.dnet.nasa.gov

Maine Space Grant Consortium

Dr. Terry Shehata
Director, Maine Space Grant Consortium
Maine Science and Technology Foundation/EPSCoR
87 Winthrop Street
Augusta , ME 04330
Tel: (207) 621-6350
Fax: (207) 621-6369
Email: shehata@mstf.org

Maryland Space Grant Consortium

Dr. Richard C. Henry, Director
The Johns Hopkins University
Department of Physics and Astronomy
Bloomberg Center for Physics and Astronomy
3400 North Charles Street
Baltimore , MD 21218-2695
Tel: (410) 516-7350
Fax: (410) 516-4109
Email: rch@pha.jhu.edu

Massachusetts Space Grant Consortium

Prof. Laurence Young
Director, Massachusetts Space Grant Consortium
Massachusetts Institute of Technology
Department of Aeronautics & Astronautics
Bldg 37, Room 207
77 Massachusetts Avenue
Cambridge , MA 02139
Tel: (617) 253-7759
Fax: (617) 253-0823
Email: lry@mit.edu

Michigan Space Grant Consortium

Dr. Roberta M. Johnson
Director, Michigan Space Grant Consortium
University of Michigan
Atmospheric, Oceanic and Space Sciences
2106 Space Research Building
Ann Arbor , MI 48109-2143
Tel: (7340 747-3430
Fax: (734) 763-0437
Email: rmjohnsn@umich.edu

Minnesota Space Grant Consortium

Dr. William L. Garrard
Director, Minnesota Space Grant Consortium
University of Minnesota
Department of Aerospace Engineering & Mechanics
107 Akerman Hall
110 Union Street Southeast
Minneapolis , MN 55455
Tel: (612) 625-9002
Fax: (612) 626-1558
Email: garrard@aem.umn.edu

Mississippi Space Grant Consortium

Dr. Michael R. Dingerson
Director, Mississippi Space Grant Consortium
University of Mississippi
125 Old Chemistry
University , MS 38677
Tel: (601) 232-5232
Fax: (601) 232-7577
Email: mdingers@olemiss.edu

Missouri Space Grant Consortium

Dr. Bruce Selberg
University of Missouri, Rolla
Executive Member
Dept. of Mechanical & Aerospace Eng. & Eng. Mech.
101 Mechanical Engineering Building
Rolla , MO 65401-0249
Tel: (314) 341-4671
Fax: 3143414607
Email: bpaul@umr.edu

Montana Space Grant Consortium

Dr. William A. Hiscock
Director, Montana Space Grant Consortium
Montana State University
Department of Physics
Engineering/Physical Sciences Building
Bozeman , MT 59717-3840
Tel: (406) 994-6170
Fax: (406) 994-4452
Email: billh@orion.physics.montana.edu

Nebraska Space Grant Consortium

Dr. Brent D. Bowen, Director
University of Nebraska, Omaha
Aviation Institute
Allwine Hall 422
6001 Dodge Street
Omaha , NE 68182-0406
Tel: (402) 554-3772
Fax: (402) 554-3781
Email: nasa@unomaha.edu

Nevada Space Grant Consortium

Dr. James V. Taranik
Director, Nevada Space Grant Consortium
Desert Research Institute
755 East Flamingo Road
Las Vegas , NV 89132-0040
Tel: (702) 895-0406

Fax: (702) 895-0496
Email: 74514.3607@compuserve.com

New Hampshire Space Grant Consortium

Dr. David Bartlett
Director, New Hampshire Space Grant Consortium
University of New Hampshire
Inst. for the Study of Earth, Oceans, & Space
Morse Hall
Durham , NH 03824-3525
Tel: (603) 862-0094
Fax: (603) 862-1915
Email: nhspacegrant@unh.edu

New Jersey Space Grant Consortium

Prof. Siva Thangam
Director, New Jersey Space Grant Consortium
Stevens Institute of Technology
Department of Mechanical Engineering
Castle Point on the Hudson
Hoboken , NJ 07030
Tel: (201) 216-5558
Fax: (201) 216-8315
Email: sthangam@stevens-tech.edu

New Mexico Space Grant Consortium

Dr. Patricia Hynes, Ph.D
Director, New Mexico Space Grant Consortium
New Mexico State University
Department of Electrical and Computer Engineering
Box 30001, Dept. SG
Las Cruces, NM 88003
Tel: (505) 646-6414
Fax: (505) 646-7791
Email: phynes@pathfinder.nmsu.edu

New York Space Grant Consortium

Dr. Yervant Terzian
Director, New York Space Grant Consortium
Cornell University
Department of Astronomy
512 Space Sciences Building
Ithaca , NY 14853-6801
Tel: (607) 255-4935
Fax: (607) 255-9817
Email: terzian@astrosun.tn.cornell.edu

North Carolina Space Grant Consortium

Dr. Fred R. DeJarnette
Director, North Carolina Space Grant Consortium
North Carolina State University
Dept. of Mechanical and Aerospace Engineering
Box 7515
1009 Capability Drive, Room 216E
Raleigh , NC 27695-7515
Tel: (919) 515-4240
Fax: (919) 515-5934
Email: dejar@ncsu.edu

North Dakota Space Grant Consortium

Dr. Charles A. Wood
Director, North Dakota Space Grant Consortium
University of North Dakota
Department of Space Studies
Center for Aerospace Sciences
Box 9008
Grand Forks , ND 58202
Tel: (701) 777-3167
Fax: (701) 777-3016
Email: cwood@badlands.nodak.edu

Ohio Space Grant Consortium

Dr. Kenneth DeWitt, Director
Ohio Aerospace Institute
Department of Chemical Engineering
University of Toledo
2801 East Bancroft Street
NH 3060
Toledo , OH 43606
Tel: (419) 530-8094
Fax: (419) 530-8086
Email: kdewitt@uoft02.utoledo.edu

Oklahoma Space Grant Consortium

Dr. Victoria Duca-Snowden
Director, Oklahoma Space Grant Consortium
University of Oklahoma
Oklahoma Climatological Survey
College of Geosciences
100 East Boyd , Suite 1210
Norman , OK 73019-0628
Tel: (405) 325-1240
Fax: (405) 325-2550
Email: vduca@ou.edu

Oregon Space Grant Consortium

Dr. Andrew C. Klein
Director, Oregon Space Grant Consortium
Oregon State University
Department of Nuclear Engineering
130 Radiation Center
Corvallis , OR 97331
Tel: (541) 737-2414 Ext: 5902
Fax: (541) 737-0480
Email: kleina@ccmail.orst.edu

Pennsylvania Space Grant Consortium

Dr. Richard Devon

Director, Pennsylvania Space Grant Consortium
Pennsylvania State University
101 South Frear
University Park , PA 16802-6004
Tel: (814) 863-7687
Fax: (814)8638286
Email: duf@psu.edu

Puerto Rico Space Grant Consortium

Dr. Juan G. Gonzalez Lagoa, Director
University of Puerto Rico, Mayaguez
Resource Center for Science and Engineering
P.O. Box 9027 College Station
Mayaguez , PR 00681
Tel: (787) 831-1022
Fax: (787) 832-4680
Email: ju_gonzalez@rumac.upr.clu.edu

Rhode Island Space Grant Consortium

Dr. Peter H. Schultz
Director, Rhode Island Space Grant Consortium
Brown University
Department of Geological Sciences
Box 1846
Providence , RI 02912
Tel: (401) 863-2417
Fax: (401) 863-3978
Email: peter_schultz@brown.edu

Rocky Mountain Space Grant Consortium

Dr. Doran J. Baker, Co-Director
Rocky Mountain Space Grant Consortium
Utah State University
College of Engineering
Engineering Building, Room EL 302
Logan , UT 84322-4140
Tel: (801) 797-3666
Fax: (801) 797-3382
Email: rmc@sdl.usu.edu

South Carolina Space Grant Consortium

Dr. Mitchell W. Colgan
Director, South Carolina Space Grant Consortium
College of Charleston
Department of Geology
66 George Street
Charleston , SC 29424
Tel: (803) 953-5463
Fax: (803) 953-5446
Email: mcolgan@jove.cofc.edu

South Dakota Space Grant Consortium

Dr. Sherry O. Farwell
Director, South Dakota Space Grant Consortium
South Dakota School of Mines and Technology
Graduate Education and Sponsored Programs
501 East St. Joseph Street
Rapid City , SD 57701-3995
Tel: (605) 394-2493
Fax: (605) 394-5360
Email: sfarwell@silver.sdsmt.edu

Tennessee Space Grant Consortium

Dr. Alvin M. Strauss
Director, Tennessee Space Grant Consortium
Vanderbilt University
Department of Mechanical Engineering
Box 1617, Station B
Nashville , TN 37235
Tel: (615) 322-2950
Fax: (615) 343-6687
Email: ams@vuse.vanderbilt.edu

Texas Space Grant Consortium

Dr. Byron Tapley
Director, Texas Space Grant Consortium
University of Texas, Austin
3925 West Braker Lane, Suite 200
Center for Space Research
Austin , TX 78759-5321
Tel: (512) 471-3583
Fax: (512) 471-7363
Email: tapley@csr.utexas.edu

Vermont Space Grant Consortium

Prof. William D. Lakin
Director, Vermont Space Grant Consortium
Department of Mathematics & Statistics
16 Colchester Avenue
Burlington , VT 05401
Tel: (802) 656-8541
Fax: (802) 656-2552
Email: lakin@emba.uvm.edu

Virginia Space Grant Consortium

Ms. Mary Sandy
Director, Virginia Space Grant Consortium
Virginia Space Grant Consortium Headquarters
2713-D Magruder Boulevard
Old Dominion University Peninsula Center
Hampton , VA 23666-1563
Tel: (757) 865-0726
Fax: (757) 865-7965
Email: msandy@pen.k12.va.us

Washington State Space Grant Consortium

Dr. Janice DeCosmo
Director, Washington Space Grant Consortium

University of Washington
Space Grant Box 351650
319 Johnson Hall
Seattle , WA 98195-1650
Tel: (206) 685-8542
Fax: (206) 685-3815
Email: janice@geophys.washington.edu

West Virginia Space Grant Consortium

Dr. Majid Jaraiedi
Director, West Virginia Space Grant Consortium
West Virginia University
College of Engineering
108A Engineering Research Building
P.O. Box 6107
Morgantown , WV 26506-6107
Tel: (304) 293-4099 Ext: 672
Fax: (304) 293-4970
Email: Jaraiedi@cemr.wvu.edu

Wisconsin Space Grant Consortium

Dr. John W. Norbury
Advisory Council Member & Institutional Rep.
University of Wisconsin, Milwaukee
Department of Physics
P.O. Box 413
Milwaukee , WI 53101-0413
Tel: (414) 229-4969
Email: norbury@csd.uwm.edu

Wyoming Space Grant Consortium

Dr. Paul E. Johnson
Director, Wyoming Space Grant Consortium
University of Wyoming
Department of Physics and Astronomy
P. O. Box 3905
University Station
Laramie , WY 82071-3905
Tel: (307) 766-6267
Fax: (307) 766-2652
Email: pjohnson@uwyo.edu

State Financial Assistance Agencies

This section provides you with a listing of state agencies that offers assistance to qualified students. These state agencies may be restricted to their legal residents. However, they may also be available to out-of-state students who will be or are attending public or private colleges or universities within the state. Followed after the state's name is the agency that may provide you with some assistance.

Alabama

Student Assistance Program
Alabama Commission on Higher Education
1 Court Square, Suite 221
Montgomery, AL 36104-3584
Tel. (205) 269-2700

Alaska

Alaska Commission on Post-Secondary Education
Box 110505
Juneau, AK 99811-0505
Tel. (907) 465-2962

Arizona

Commission on Post-Secondary Education
2020 N. Central Ave., Suite 275
Phoenix, AZ 85004
Tel. (602) 229-2593

Arkansas

Department of Higher Education
114 East Capitol
Little Rock, AR 72201-3818
Tel. (501) 324-9300

California

Student Aid Commission
Box 510845
Sacramento, CA 94245-0845
Tel. (916) 445-0880 or 322-9267

Colorado

Colorado Commission on Higher Education
1300 Broadway, 2nd Flr.
Denver, CO 80203
Tel. (303) 866-2723

Connecticut

Student Financial Assistance Commission
Department of Higher Education
61 Woodland St.
Hartford, CT 06105-2391
Tel. (203) 566-2618

Delaware

Delaware Post-Secondary Education Commission
State Office Building
820 N. French St., 4th floor
Wilmington, DE 19801
Tel. (302) 571-3240

District of Columbia

Office of Post-Secondary Education
2100 Martin Luther King Avenue, S.E., Ste 401
Washington, DC 20020
Tel. (202) 727-3685/3688

Florida

Student Financial Assistance Commission
Florida Department of Education Center
1344 Florida Education Center
Tallahassee, FL 32399-0400
State Aid: (904) 488-1034 GSL:(904) 488-4095

Georgia

Georgia Student Finance Authority
2082 East Exchange Place, Suite 200
Tucker, GA 30084
Tel. (404) 493-5452

Hawaii

State Post-secondary Education Commission
Bachman Hall, Rm 209
University of Hawaii
2444 Dole St.
Honolulu, HI 96822
Tel. (808) 956-8213

Idaho

State Board of Education
650 West State St., Rm 307
Boise, ID 83720
Tel. (208) 334-2270

Illinois

State Scholarship Commission
1755 Lake Cook Road
Deerfield, IL 60015-5209
Tel. (708) 948-8550

Indiana

State sdtudent Assistance Commission
150 West Market Street, Ste 500
Indianapolis, IN 46204-2811
Tel. (317) 232-2350

Iowa

College Aid Commission
201 Jewett Building
914 Grand Ave.
Des Moines, IA 50309-2824
Tel. (515) 281-3501

Kansas

Board of Regents, State of Kansas
Suite 1410, 700 S.W. Harrison
Topeka, KS 66603-3760
Tel. (913) 296-3517

Kentucky

Higher Education Assistance Authority
1050 U.S. 127 South, Suite 102
Frankfort, KY 40601
Tel. (502) 564-4928

Louisiana

Governor's Special Commission on Education Services
PO Box 44127
Capitol Station
Baton Rouge, LA 70804
Tel. (504) 922-1038

Office of Student Financial Assistance
P.O. Box 91202
Baton Rouge, LA 70821-9202
Tel. (504) 922-1150

Maine

Finance Authority of Maine
Education Assistance Division
State House Station #119
One Weston Court
Augusta, ME 04333
Tel. (207) 287-2183

Maryland

State Scholarship Board
2100 Guilford Ave., Rm 207
Baltimore, MD 21218
Tel. (301) 333-6420

Massachusetts

Board of Regents of Higher Education
Scholarship Office
330 Stuart St.
Boston, MA 02116
Tel. (617) 727-9420

Michigan

Michigan Higher Education Assistance Authority
PO Box 30008
Lansing, MI 48909-7508
Tel. (517) 373-3394 or 3399

Minnesota

Minnesota Higher Education Coordinating Board
Capitol Square Building, Suite 400
550 Cedar St.
Saint Paul, MN 55101
Tel. (612) 296-3974/9657 x3034

Mississippi

Postsecondary Education Financial Assistance Board
3825 Ridgewood Road
Jackson, MS 39221-6453
Tel. (601) 982-6570 or 6661

Missouri

Corrdinating Board for Higher Education
101 Adams Street
Jefferson City, MO 65101-3059
Tel. (314) 751-2361

Montana

Montana Guaranteed Student Loan Program
2500 Broadway
Helena, MT 59604
Tel. (406) 444-6594

Nebraska

Nebraska Coordinating Commission for Post-secondary Education
P.O. Box 95005
Lincoln, NE 68509
Tel. (402) 471-2847

Nevada

Financial Aid Office
University of Nevada, Reno
Rm 200 TSSC
Reno, NV 89557
Tel. (702) 784-4666

Nevada Department of Education
Capitol Complex
400 W. King Street
Carson City, NV 89710
Tel. (702)687-5915

New Hampshire

New Hampshire Education Commission
2 Industrial Park Drive
Concord, NH 03301-8512
Tel. (603) 271-2555

New Jersey

Department of Higher Education
Office of Student Assistance
4 Quakerbridge Plaza CN 540
Trenton, NJ 08625
Tel. (609) 588-3268

New Mexico

Commission on Higher Education
1068 Cerrillos Road
Santa Fe, NM 87501-4295
Tel. (505) 827-7383

New York

Higher Education Service Corporation
One Commerce Plaza
Albany, NY 12255
Tel. (518) 473-0431 or 7087

North Carolina

State Education Assistance Authority
PO Box 2688
Chapel Hill, NC 27515-2688
Tel. (919) 549-8614

North Dakota

Student Financial Assistance Program
600 East Boulevard
Bismarck, ND 58505
Tel. (701) 224-4114

Ohio

Ohio Board of Regents
30 East Broad St., 36th Flr
Columbus, OH 43266-0417
Tel. (614)466-1191

Oklahoma

Oklahoma State Regents for Higher Education
P.O. Box 3020
Oklahoma City, OK 73101-3020
Tel. (405) 525-4356

Oregon

State Scholarship Commission
1500 Valley River Drive, Suite 100
Eugene, OR 97401
Tel. (503) 346-4166

Pennsylvania

Higher Education Assistance Agency
660 Boas St.
Harrisburg, PA 17102
Tel. (717) 257-2800 or 3300

Rhode Island

Higher Education Assistance Authority
560 Jefferson Boulevard
Warwick, RI 02886
Tel. (401) 277-2050 or (800) 922-9855

South Carolina

South Carolina Tuition Grants Agency
PO Box 12159
Columbia, SC 29211
Tel. (803) 734-1200

South Dakota

Office of the Secretary
Department of education & Cultural Affairs
700 Governor's Dr.
Pierre, SD 57501-2291
Tel. (605) 773-3134

Tennessee

Tennessee Student Assistance Corporation
404 James Robertson Parkway,
1950 Parkway Tower
Nashville, TN 37243-0820
Tel. (615) 741-1346 or (800) 342-1663

Texas

Texas Higher Education Coordinating Board
7745 Chevy Chase Drive, Capitol Station
Austin, TX 78752
Tel. (512) 483-6340

Utah

Utah State Board of Regents
335 West North Temple
3 Triad Center, Suite 550
Salt Lake City, UT 84180-1205
Tel. (801) 538-5247

Vermont

Vermont Student Assistance Corporation
Champlain Mill, Box 2000
Winooski, VT 05404-2000
Tel. (802) 655-9602

Virginia

Council of Higher Education
James Monroe Building
101 North 14th St.
Richmond, VA 23219

Tel. (804) 225-2623

Washington

Higher Education Coordinating Board
917 Lakeridge Way, GV-11
Olympia, WA 98504
Tel. (206) 753-2210

Wisconsin

State of Wisconsin Higher Education Aids Board
PO Box 7885
Madison, WI 53707-7885
Tel. (606) 266-1660

Wyoming

University of Wyoming
Student Financial Aid
PO Box 3335, University Station
Laramie, WY 82701

Wyoming Community College Commission
122 West 25th Street
Cheyenne, WY 82002
Tel. (307) 777-7227

Guam

Student Financial Assistance
Universaity of Guam, UOG State
Mangilao, Guam 96923
Tel. (671) 734-4469

Puerto Rico

Council on Higher Education
Box 23305, UPR Station
Rio Piedras, Puerto Rico 00931
Tel. (809) 758-3350

Virgin Islands

Board of Education
Commandant Gade
O.V. No. 11
St. Thomas, Virgin Islands 00801
Tel. (809) 774-4546

THIS PAGE INTENTIONALLY LEFT BLANK

Glossary

The process of awarding student financial aid has grown more complex over the years, and as a result has developed its own vocabulary. To help reduce confusion for students and parents, this section presents common definitions for many of the words used by financial aid adminstrators and scholarship sponsors.

Academic Year
The period during which school is in session, consisting of at least 30 weeks of instructional time. The school year typically runs from the beginning of September through the end of May at most colleges and universities.

Accrual Date
The date on which interest charges begin to accrue.

Accrued Interest
Interest on a loan that accumulates and is to be paid in installments at a later time usually when the principal becomes due. Accrued interest may be compounded or simple.

Advanced Placement Test (AP)
The Advanced Placement tests are used to earn credit for college subjects while in high school.

American College Test (ACT)
A standardized college entrance examinations used in the United States.

Amortization
The process of slowly repaying a loan over an extended period of time through monthly installments of principal and interest.

Appeal
A formal request to have a financial aid administrator review your financial aid eligibility and possibly use Professional Judgment to readjust the figures.

Assets
The amount a family has in savings and investments. This includes savings and checking accounts; a business; a farm or other real estate; and stocks, bonds, and trust funds.

Associate Degree
A two-year undergraduate college degree.

Award Year
An academic year in which a student will recieve or request financial aid.

Bachelor's Degree
A four-year undergraduate degree offered by colleges and universities.

Balloon Payment
A larger payment used to pay off the outstanding balance of a loan without penalty.

Billing Servicer
A company that manages the billing and collection of loans for lenders.

Borrower
A person or group that receives a loan

Bursar's Office
A university's office that is responsible for the billing and collection of university charges to students.

Business/Farm Supplement
An additional financial aid form required by some colleges for parents and students who own a business or farm.

Cancellation
Some loan programs provide for cancellation of the loan under certain circumstances. The borrower may be eligible for cancellation of all or part of the balance of his/her educational loans.

Capitalization
The process of increasing the size of the loan by adding unpaid interest charges to the principal.

Citizenship/Eligibility For Aid
To be eligible to receive federally funded college aid, a student must be one of the following: 1. a United States citizen, 2. a non-citizen national, 3. a permanent resident with an I-151 or I-551 without conditions, 4. a holder of an I-94 showing one of the following designations: "Refugee," "Asylum Granted," "Indefinite Parole" and/or "Humanitarian Parole," "Cuban-Haitian Entrant, Status Pending," "Conditional Entrant" (valid if issued before April 1, 1980), 5. a participant in a suspension of deportation case pending before Congress Individuals in the U.S. on an F1 or F2 visa only or on a J1 or J2 exchange visa only cannot get federal aid.

Collateral
Something of value used as security for a loan.

Collection Agency
An outside agency used by the lender or guarantee agency to recover defaulted loans.

College Scholarship Service
The College Board and one of the agencies that processes financial aid information and applications.

Compounded Interest
interest that is paid on both the principal balance of the loan and on any accrued (unpaid) interest, resulting in a new principal balance which will have a new interest assessment.

Consolidation Loan
This loan allows a borrower to combine several educational loans into one new loan. This sometimes results in a lower interest rate. Such loans often reduce the size of the monthly payment by extending the term (up to 30 years depending on the loan amount) of the loan and allowing a single monthly payment, consolidation can make loan repayment easier for some borrowers.

Cosigner
A second creditworthy person who signs a promissory note with a borrower who does not have collateral or a good credit history. The cosigner guarantees that the loan will be repaid if the borrower fails to make payments.

Cost Of Education (Or Cost Of Attendance)
The total amount it will cost a student to attend college for a year, including tuition and fees; housing and food for the period of enrollment; books and supplies; and miscellaneous expenses (travel costs, flight fees, lab fees etc.). Other expenses may be added at the discretion of a college's financial aid administrator.

Cooperative Education
A program offered by many instutions in which students alternate periods of enrollment with periods of employment (some provide variuos pay rates). This program may extend the regular baccalaureate degree program to five years. This program combines classroom study with actual work experience.

Default
Failure to repay a student loan installment on time (failure to pay several regular installments, i.e. payments overdue by 180 days) according to the terms and conditions agreed to when you signed a promissory note. If you default on a loan, the university, the holder of the loan, and the government can take legal action to recover the money, including garnishing your wages. Defaulting on a government loan will make you ineligible for future federal financial aid. Defaults are recorded on permanent credit records and may result in prosecution and/or loss of future borrowing possibilities.

Deferment
A borrower is allowed to postpone repaying the loan. Deferments are available while borrowers are in school at least half time, enrolled in a graduate fellowship program or rehabilitation training program, and during periods of unemployment or economic hardship. Other loan programs allow the student to defer the interest payments by capitalizing the interest. Other deferments may be available depending on when and what you borrowed. This benefit is generally characteristic of federal and state guaranteed student loans. Contact your lender for additional details.

Delinquent
The borrower fails to make a payment on time, the borrower is now considered delinquent and late fees may be charged. If several installments are not paid, the loan goes into default.

Dependent Student
A student claimed as a dependent member of household for federal income tax purposes.

Direct Loan
Direct Loan is a new federal program where the school becomes the lending agency and manages the funds directly, with the federal government acting as the guarantee agency. Not all schools currently participate in this program. The William D. Ford Federal Direct Loan Program also known as the Direct Loan Program.

Disbursement
The date on which the loan funds are released to the university for payment. Disbursements for most loans are made in equal multiple installments, and made co-payable to the borrower and the school.

Discharge
To release the borrower from his/her obligation to repay the loan

Disclosure Statement
This statement provides the borrower with information about the actual cost of the loan, including the interest rate, origination, insurance, and loan fees, and any other kinds of finance charges. Lenders must provide the borrower with a disclosure statement before issuing a loan.

Discretionary Income
Income that is available to a person or family after all financial obligations, including taxes, have been accounted for.

Due Diligence
The federal government requires the lender of the loan to make numerous attempts to contact the borrower by telephone and mail, if the borrower failed to make payments according to the terms of the promissory note. This is to encourage the borrower to repay the loan and make arrangements to resolve the delinquency.

Eligible Noncitizen
A financial aid applicant who is not a U.S. citizen but is eligible to receive federal Title IV aid because he/she is a permanent resident, noncitizen national, or a resident of the Trust Territory of the Pacific Islands or Micronesia.

Estimated Family Contribution (EFC)
The amount of money the financial aid office expects the family to be able to contribute to the student's education. The EFC is calculated according to a formula established by Congress. The difference

between the COA and the EFC is the student's financial need.

Electronic Funds Transfer (EFT)
This process is used by lenders to wire funds directly to participating schools without requiring an intermediate check for the student to endorse.

Enrollment Status
The number of credits a student is taking per semester or quarter. The status is diveded into two caterogies: full-time (usually 12 credits or more) or part-time (usually 6 credits) status. A student must be enrolled at least half-time or full-time to qualify for financial aid.

Entrance Interview
A required counseling session at which a financial aid officer, must inform the student borrowers about their rights and responsibilities.

Exit Interview
Students with educational loans are required to meet with a financial aid officer before they graduate. During this exit interview, an adminstrator reviews the repayment terms of the loan and the repayment schedule with the student.

Federal Direct Loan
The federal government is the lender for a group of federal loan programs. Included in these programs are government-subsidized loans for students and unsubsidized loans for both students and parents.

Federal Education Loan Programs
A bank, savings and loan, credit union, or other private organization is the lender for a group of federal loan programs. Included in these programs are government-subsidized loans for students and unsubsidized loans for both students and parents.

Federal Methodology
A standard method of calculating how much a family should be expected to contribute toward college costs. All the federal funds are awarded based on this need analysis formula.

Federal Pell Grant
Federal grant awarded to undergraduate students based on need.

Federal Perkins Loan
A 5% loan funded by the government which is awarded by colleges to both undergraduate and graduate students.

Federal Plus Loan
A nonsubsidized loan program for parents of undergraduate students under the Federal Education Loan Program umbrella.

Federal Stafford Loan
A Federal Education Loan Program for students. Stafford Loans can be either government-subsidized, in which case the government pays any interest while the borrower is attending college, or unsubsidized, in which case interest begins to accrue when the loan is made.

Federal State Student Incentive Grant
Awards made as part of a state grant program utilizing both federal funds and state funds.

Federal Supplemental Educational Opportunity Grant (FSEOG)
A federal grant awarded by colleges to the most needy undergraduate students as determined by the federal need analysis formula.

Federal Work-Study Program (FWSP)
A federal, need-based financial aid program through which eligible students can earn a portion of their college expenses. Work-study awards are made by colleges, but a portion of the funding comes from the federal government. Essentially FWSP pays a portion of the student's salary, making it cheaper for departments and businesses to hire the student.

Federal Family Education Loan Program (FFELP)
This program includes the Federal Stafford Loans and the PLUS loans.

Fellowship
A form of financial aid given to graduate students to help support their education. Some fellowships include a tuition waiver or a payment to the university in lieu of tuition. Most fellowships include a stipend to cover reasonable living expenses.

Financial Aid Administrator (FAA)
A FAA is an university employee who is involved in the financial aid process.

Financial Aid Award Letter
Written notification to an applicant from a college that details how much and which types of financial aid are being offered if the applicant enrolls.

Financial Aid Package
The total amount of financial aid a student receives. This is a complete collection of grants, scholarships, loans, and work-study program offered to a student to financially assist them to attend the university for one school year.

Financial Aid Profile
A financial aid application developed by the College Scholarship Service that many colleges use to determine aid given from their own institutional funds.

Financial Aid Transcript (FAT)
A record of any financial aid a student has received at a given institution. To be eligible for federal financial aid programs, students must submit this transcript from all previously attended postsecondary institutions, regardless of whether aid was received.

Financial Need
The difference between a college's cost of attendance and the Expected Family Contribution) as calculated by the need analysis methodology.

Free Application For Federal Student Aid (FAFSA)
The official application students must use to apply for federal aid.

Financial Aid Office (FAO)
The college's office that handles all matters affecting financial aid.

Fixed Interest
A rate of interest that is set at the time a loan is negotiated and that remains constant over the life of the loan.

Forbearance
An authorized period of time during which the lender agrees to temporarily postpone a borrower's principal repayment obligation. Interest continues to accrue and usually must be paid during the forbearance period. Forbearance may be granted at the lender's discretion .

Garnish
The process of withholding a portion of a borrower's wages to repay his/her loan, usually without their consent.

Gift Aid
Grant and scholarship funds given as financial aid that does not have to be repaid.

Grace Period
A short time period after graduation or no longer in school during which the borrower is not required to begin repayment. The typical grace period is six or nine months. Some loan programs have no grace period.

Grade Point Average (GPA)
An average of a student's grades. Some schools uses the 4.0 scale (4.0 = A, 3.0 = B, 2.0 = C, 1.0 = D) and others uses the 5.0 scale.

Grant
A type of financial aid award based on need or merit that is not repaid by the student.

Gross Income
A family's or individual's total income before deductions.

Guarantee Agency
A guarantee agency agrees to pay back a loan due to a borrower's default, death, disability, or bankruptcy. The federal government sets loan limits and interest rate, but each state is free to set its own additional limits within the federal guidelines. Each state has a different guarantee agency that administers the federal Stafford and Plus loans for students in that state. A guarantee fee is a small percentage of the loan that is paid to the guarantee agency as a form of insurance against default. For the name, address, and telephone number of your state's guarantee agency, call the Federal Student Aid Information Center at 1-800-433-3243 (1-800-4-FED-AID).

Guarantee Fee
An insurance fee, deducted from the borrower's loan proceeds prior to disbursement, that the guarantee agency charges a lender. The fee for a Federal Stafford or PLUS Loan is1%. By law the fee cannot exceed 3% of the loan amount.

Independent Student
A student who reports only his or her own income (and that of a spouse, if relevant) when applying for federal financial aid. Students who will be 24 or older by December 31, 1998, will automatically be considered "independent" for 1998-1999. Students who are under 24 will be considered independent if they are: 1. married and not claimed as a dependent on their parents' 1998 federal income tax return; 2. the supporter of a legal dependent other than a spouse; 3. a veteran of the U.S. Armed Forces; 4. an orphan or ward of the court; 4. classified as independent by a college's financial aid; 5. administrator because of other unusual circumstances; 6. a graduate or professional student

Indirect Costs
All the non-tuition-related costs associated with attending college, including room, board, transportation, medical, and personal expenses.

In-State Student
The legal resident of state, and the student is eligible for reduced in-state student tuition at a public college or universitie in the state. Certain requirments must be met.

Institutional Methodology
A standard method of determining a student's or family's ability to pay for college used by individual colleges in awarding their own institutional funds for financial aid. However, colleges must use the Federal Methodology in awarding any federal funds.

Insurance Fee
A fee charged to guarantee student loans against loss through default. The amount charged is usually deducted from the disbursement of the principal.

Interest
A fee charged to the borrower for the use of borrowed money from the lender. Interest is calculated as a percentage of the principal loan amount. The percentage rate may be fixed for the life of the loan, or it may be variable, depending on the terms of the loan. As of October 1, 1992, all federal education loans made to new borrowers have variable interest rates.

Interest Subsidy
Interest payments made by the federal government to the lender of a Subsidized Federal Stafford or Direct Loan while the borrower is enrolled at least half-time or is in a grace period.

Internal Revenue Service (IRS)
The federal agency responsible for collecting income taxes. Student aid applications are often verified by their family's IRS forms.

Internship
A part-time job during the academic year or the summer months. A student receives supervised work experience in a his/her career field. Some internships provide the student with a mentor and stipend.

Lender
A financial institution (bank, savings and loan, or credit union) that provides the money for students and parents to borrow educational loans. The money borrowed usually have conditions that the money be returned with an interest charge. Some schools are also lenders.

Loan Disclosure Statement
A document that shows the amount of a loan; where, when, and what repayments must be made; the interest rate; and the cost of borrowing that loan.

Merit-Based Aid
Any form of financial aid awarded on the basis of personal achievement or individual characteristics without reference to financial need.

Need
The difference between the COA and the EFC is the student's financial need. The remaining amount is considered student need. The financial aid package is often based on the amount of financial need. The process of determining a student's need is known as the need analysis.

Needs Analysis
A process of reviewing a student's aid application to determine the amount of financial aid a student is eligible for. Completing a needs analysis form is the required first step in applying for most types of financial aid.

New Borrower
A borrower who has no outstanding (unpaid) loan balances on the date he/she signs the promissory note for a specific educational loan. New borrowers may be subject to different regulations than borrowers who have existing loan balances.

Origination Fee
A processing fee charged to a borrower which is deducted from the loan to pay part of the loan's adminstrative costs. This fee is usually subtracted from the amount of a loan.

Outside Scholarship
An award that recieved from sources (private scholarship, companies, foundations, etc.) other than the school and the federal or state government.

Out-of-State Student
A student has not met the legal residency requirements for the state. An out-of -state student is often charged a higher tuition fee at a public college or university in the state.

Overaward
A student's family contribution plus any financial aid awarded exceeds the cost of attendance at a given college. Overawards may result when a student's enrollment status changes or when additional resources (such as a private scholarship) become available to a student.

Pell Grant
The Pell grant is a federal grant that provides funds of up to $2,500 based on the student's financial need.

Perkins Loan
Formerly the National Direct Student Loan Program. The Perkins Loan allows students to borrow up to $3,000/year (5 year max) for undergraduate school and $5,000/year for graduate school (6 year max).

Prime Rate
The fluctuating interest rate that banks charge to their best business customers.

Principal
The amount borrowed. Interest is charged on this amount as a percentage of the principal, and guaranty and origination fees will be deducted prior to disbursement.

Prepayment
Paying off all or part of a loan before it is due.

Professional Judgment
The legal authority of financial aid administrators to change a calculated Expected Family Contribution or any of the elements used in the calculation based on additional information or individual circumstances that would lead to a more accurate assessment of a family's financial condition.

Profile
The secondary application for financial aid processed through the College Scholarship Service (CSS). This must be completed by all students who wish to receive aid other than Federal Stafford Loans.

Promisory Note
This is binding legal document signed by the student borrower before loan funds are disbursed by the lender. The promisory note states the terms and conditions of the loan, including repayment schedule, interest rate, deferment policy, and

cancellations. The student should keep this document until the loan has been repaid.

Parent Loans for Undergraduate Students (PLUS)
A federal loans available to parents of undergraduate students to help finance the student's education. Parents may borrow up to the full cost of their children's education, less the amount of any other financial aid received.

Repayment Schedule
A plan that discloses the borrower's monthly payment, interest rate, total repayment obligation, due dates and length of time for repaying the loan.

Repayment Term
A number of installments or years the borrower is required to make payments on his/her loans.

Research Assistantship (RA)
A form of financial aid given to graduate students to help support their education. Research assistantships usually provide the graduate student with a waiver of all or part of tuition, plus a small stipend for living expenses. As the name implies, an RA is required to perform research duties. Sometimes these duties are strongly tied to the student's eventual thesis topic.

Renewable Scholarships
A scholarship that is awarded for more than one year. Usually the student must maintain certain academic standards or grade point average to be eligible for subsequent years of the award. Renewable scholarships may require the student to reapply for the scholarship each year; others may require a report on the student's progress in his/her career field.

Satisfactory Academic Progress (SAP)
The level of academic achievement expected of a student in order to continue to receive financial aid. If a student fails to maintain an academic standing consistent with the school's SAP policy, they are unlikely to meet the school's graduation requirements.

Student Aid Report (SAR)
A SAR is the sent to the student after filing a FAFSA. The SAR summarizes the information included in the FAFSA and must be provided to your school's FAO. The SAR will also indicate the amount of Pell Grant eligibility, if any.

Scholarship
A form of financial aid given to undergraduate students to help pay for their education. Most scholarships are restricted to paying all or part of tuition expenses, though some scholarships also cover room and board.

Secondary Market
Loans are often bought and sold on the secondary market. Institutions that buy loans from lenders, usually at a discount. If a loan is sold, the secondary market is responsible for managing and servicing it. The terms of your loan do not change when it is sold to another lender.

Self-Help Aid
Funds from jobs and from loan programs, such as the Federal Perkins Loan, Federal Stafford and Direct Loans, and Federal Work-Study Program.

Self-Help Expectation
The principle that students have an obligation to help pay for a portion of their own education. The expected amount of self-help is usually included in the analysis of a student's resources.

Simple Interest
Interest computed only on the original amount of a loan.

Simplified Needs Test
A formula used in the Federal Methodology for families whose Adjusted Gross Income (AGI) is less than $50,000 and who file either the 1040A or 1040EZ IRS forms. In this formula, a family's assets are not included.

Stafford Loans
Stafford Loans are federal loans that come in two forms, subsidized and unsubsidized. Subsidized loans are based on need; unsubsidized loans aren't. The Subsidized Stafford Loan was formerly known as the Guaranteed Student Loan (GSL). Undergraduates may borrow up to $23,000 ($2,625 during the freshman year, $3,500 during the sophomore year, and $5,500 during each subsequent year) and graduate students up to $65,500 including any undergraduate Stafford loans ($8,500 per year).

Statement Of Educational Purpose
A separate form, or a statement on the FAFSA, that all students must sign in order to receive federal student aid. By signing you agree that: 1. You are to use your student aid only for education-related expenses, 2. You have complied with Selective Service requirements by registering with the Selective Service or indicating the reason why you are not required to register.

Supplemental Education Opportunity Grant (SEOG)
The SEOG is a federal grant program for undergraduate students and first baccalaureate degree only with exceptional need. SEOG grants are awarded by the school's financial aid office, and provide up to $4,000 per year. Eligibility is based on the federal methodology using the information provided on the FAFSA. Priority is given to students who are eligible for Pell Grants, meet filing deadlines and who demonstrate the greatest historical need.

Subsidized Loan
A need-based loan on which the interest is paid by the federal government during the in-school, grace, and deferment periods.

Teaching Assistantship (TA)
A form of financial aid given to graduate students to help support their education. Teaching assistantships usually provide the graduate student with a waiver of all or part of tuition, plus a small stipend for living expenses. As the name implies, a TA is required to perform teaching duties.

Term
The term of a loan is the number of years (or months) during which the loan is to be repaid.

Unmet Need
The student's financial aid package and the family contribution does not cover the costs of attending a particular college, the difference is called the Unmet Need.

Unsubsidized Loan
An unsubsidized loan is a loan for which the government does not pay the interest. The borrower is responsible for the interest on an unsubsidized loan from the date the loan is disbursed.

Variable Interest
The rate of interest that changes during the life of a loan on a regular basis and is generally tied to an index. Some student and parent loan programs have variable interest rates that change annually based on the one-year Treasury Bill rate.

Verification
Verification is a review process in which the FAO determines the accuracy of the information provided on the student's financial aid application. During verification additional information from the student, a spouse, and the parents is used to confirm previously submitted documentation for the amounts listed (or not listed) on the financial aid application.

Appendix

Timeline for High School Students

This timeline maybe used throughout your college years. You may have to make some changes from year to year.

Junior Year of High School

- Research and apply for public and private scholarships, grants and other sources of financial aid from organizations, foundations, companies, individuals, schools, etc. (refer to page 35 for a list of other possible opportunities. Use the notes located in the appendix section to write down the schools you are applying.).
- Your research should begin with your High School guidance counselors, teachers, libraries, and book stores
- Create a listing of scholarships sponsors you intend to write for more information.
- Write or call the schools (colleges, universities, academies, etc.) for catalogs you are interested in attending
- Review all the institution's financial aid policies and applications
  - You may be eligible to receive assistance from the state and federal government and the school's financial aid office
- If you have any questions, call and speak with a financial aid officer at the school(s) to which you are applying for admission
- Seek employment of a summer job and begin saving money for tuition, books, lab fees, personal items, travel, etc.
- Take the Preliminary Scholastic Assessment Test (PSAT) or Preliminary ACT (PACT) for practice.
  - Practice exam(s) will help improve grade chances of a better score
  - If you achieve a good score on the actual SAT and/or ACT, you may qualify for a Merit Scholarship
  - A description of Merit Scholarship is on page 17
  - One of the admission requirements for many institutions is achieving a good score

SUMMER

- Review your scholarship sponsor list and check-off those sponsors that sent you an application or any information in regards to the scholarship offered.
  - If you have not received anything from the sponsor(s), send another request letter or call.
- Make copies of all the original applications you received and use them as draft copies.
  - Start filling out the copied application(s). Do Not write on the original application for now. make all your mistakes on the copies.
  - Make your corrections and any changes on the draft. Once you have everything corrected on your draft. Have someone proof read your draft.
  - After your draft has been proof read, and reread. Have someone type all the information onto the original application. Once again, proof read the typed application for errors. Remember: professionalism and neatness.
  - Try to complete the application accurately and as son as possible.
- If you have any questions, contact the scholarship sponsor(s).
- Remember to save money from your summer job.

Senior Year of High School

FALL

- ✈ Your applications should be completed at this time.
- ✈ Make copies and submit all applications before the deadlines.
  - Call a week after mailing your application to make sure the sponsor received it.
- ✈ Request a Free Application for Student Financial Aid (FAFSA) by calling 1-800-4-FED-AID or visit the website at http://www.fafsa.ed.gov
  - This free application may be available at your counselor's office, library, or institution you are seeking admission too.
- ✈ Parents should start to gather their income tax information early to expedite the FAFSA completion

December (month)

- ✈ Your FAFSA should be completed, along with any other financial applications
- ✈ Make copies and mail the FAFSA as soon as possible - After January 1
  - Be sure to keep all of your copies in a neat and organized folder.
- ✈ Remember financial aid based on need are awarded to students with completed FAFSA applications. So, Don't Wait!

4 Weeks After Mailing your FAFSA

- ✈ You should receive the Student Aid Report (SAR)
  - Make sure you read all information and instructions.
  - Write the date you received the SAR Date(s): __________, __________
- ✈ If your SAR has to be corrected for incomplete or inaccurate information, make sure you and your parent(s) follow all of the instruction carefully.
  - Try to complete the form(s), as soon as possible.
  - Once corrected, make another copy and mail it back to the address located on the forms
- ✈ If you have further questions, contact your institution's financial aid office.
- ✈ Once you receive the SAR and there are no errors or corrections to be filed, call your institution's financial aid office. Check with them to see if they need a copy of the SAR.

EARLY SPRING

✈ Expect to receive a letter or telephone call from the scholarship sponsor(s). If you have not received anything, call the sponsor and inquire about the status of your application.

March - July

✈ Review all the Financial Aid Packages offered by the schools to which you were accepted.
✈ Document the dates of each aid package you received.

_______________, _______________, _______________, _______________

✈ Review each award and expected family and student contribution. Then review your actual family and student contribution. There may be a big difference between the expect contribution and what your family can actually contribute.
✈ Compare the financial aid package of each school and compare the types of aid offered.
✈ After reviewing and comparing, accept all or parts of the financial aid awards. read all instructions carefully.
✈Sign, copy, and return the financial aid packages/letter to the school. Document the school name(s) and date(s) of each award package you mailed.

__________________/_________________, __________________/_________________

__________________/_________________, __________________/_________________

✈ Follow all the Financial Aid Office's (FAO) Procedures, If you or your parents are applying for a student or parent loan. Note Lender Code (if applicable): _______________
✈ Contact your FAO to check on the status of the loans
✈ Contact the admissions and FAO of the schools you have decided to decline enrollment or financial aid.
✈ Look for a summer job or internship / cooperative education program

Winning a Scholarship

✈ All public or private scholarships must be reported to the school's FAO.

Congratulations on Your High School Graduation

Important Note:

You must reapply for Financial Aid each year you are attending school to receive aid from the government and/or your school. Remember the deadline dates!

Good Luck!

Qualifications / Achievements / Skills Form

Certification / Licenses / Special Training

✈ ______________________ ✈ ______________________
✈ ______________________ ✈ ______________________
✈ ______________________ ✈ ______________________
✈ ______________________ ✈ ______________________

Examples: Private Pilot Certificate, Lifeguard, Cardio-pulmonary Resuscitation (CPR) certification

Clubs / Memberships / Organizations

✈ ______________________ ✈ ______________________
✈ ______________________ ✈ ______________________
✈ ______________________ ✈ ______________________
✈ ______________________ ✈ ______________________

Examples: Alpha Eta Rho, Women in Aviation, Big Brother & Sisters, Scouting, Fraternity / Sorority

Technical Skills / Qualifications

✈ ______________________ ✈ ______________________
✈ ______________________ ✈ ______________________
✈ ______________________ ✈ ______________________
✈ ______________________ ✈ ______________________

Examples: Computer Programmer, Windows 95, Excel, Power Point, typing

Personal Interest / Hobbies / Sports

✈ ______________________ ✈ ______________________
✈ ______________________ ✈ ______________________
✈ ______________________ ✈ ______________________
✈ ______________________ ✈ ______________________

Examples: Violin, painting, viewing art, camping, basketball, reading, golf, tennis, travel

Charity / Volunteer Work

✈ ______________________ ✈ ______________________
✈ ______________________ ✈ ______________________
✈ ______________________ ✈ ______________________
✈ ______________________ ✈ ______________________

Examples: Young Eagles Volunteer, United Way Walk-a-Thon, High School Tutor

Work Experience

✈ ______________________ ✈ ______________________
✈ ______________________ ✈ ______________________
✈ ______________________ ✈ ______________________
✈ ______________________ ✈ ______________________

Examples: Ramp operator, store clerk, cashier, paper route

Qualifications / Achievements / Skills Form
Part 2

Personal Strengths

✈ ____________________ ✈ ____________________
✈ ____________________ ✈ ____________________
✈ ____________________ ✈ ____________________
✈ ____________________ ✈ ____________________

Personal Weaknesses

✈ ____________________ ✈ ____________________
✈ ____________________ ✈ ____________________
✈ ____________________ ✈ ____________________
✈ ____________________ ✈ ____________________

Long-Term and Short-Term Goals

✈ ____________________ ✈ ____________________
✈ ____________________ ✈ ____________________
✈ ____________________ ✈ ____________________
✈ ____________________ ✈ ____________________

Scholarship Track Form™
Flight Time Publishing

Scholarship Name: ____________________

Contact Person: ____________________

Deadline: ____________________

Page No./Index No.: __________ / __________

Request Letter: __________, __________

Application Received: ____________________

Application Completed/Mailed:

__________ / __________

- __ Application Complete
- __ Official Transcript
- __ Letter(s) of Recommendation
- __ Essay(s)
- __ Reference Sheet
- __ Parents Tax Returns
- __ Return Envelope
- __ Application Signed & Dated
- __ Make Copies
- __ Other: ____________________

Special Notes: ____________________

Results: ____________________

Scholarship Track Form™
Flight Time Publishing

Scholarship Name: ____________________

Contact Person: ____________________

Deadline: ____________________

Page No./Index No.: __________ / __________

Request Letter: __________, __________

Application Received: ____________________

Application Completed/Mailed:

__________ / __________

- __ Application Complete
- __ Official Transcript
- __ Letter(s) of Recommendation
- __ Essay(s)
- __ Reference Sheet
- __ Parents Tax Returns
- __ Return Envelope
- __ Application Signed & Dated
- __ Make Copies
- __ Other: ____________________

Special Notes: ____________________

Results: ____________________

Scholarship Track Form™
Flight Time Publishing

Scholarship Name: ____________________

Contact Person: ____________________

Deadline: ____________________

Page No./Index No.: __________ / __________

Request Letter: __________, __________

Application Received: ____________________

Application Completed/Mailed:

__________ / __________

- __ Application Complete
- __ Official Transcript
- __ Letter(s) of Recommendation
- __ Essay(s)
- __ Reference Sheet
- __ Parents Tax Returns
- __ Return Envelope
- __ Application Signed & Dated
- __ Make Copies
- __ Other: ____________________

Special Notes: ____________________

Results: ____________________

Scholarship Track Form™
Flight Time Publishing

Scholarship Name: ____________________

Contact Person: ____________________

Deadline: ____________________

Page No./Index No.: __________ / __________

Request Letter: __________, __________

Application Received: ____________________

Application Completed/Mailed:

__________ / __________

- __ Application Complete
- __ Official Transcript
- __ Letter(s) of Recommendation
- __ Essay(s)
- __ Reference Sheet
- __ Parents Tax Returns
- __ Return Envelope
- __ Application Signed & Dated
- __ Make Copies
- __ Other: ____________________

Special Notes: ____________________

Results: ____________________

Scholarship Track Form™
Flight Time Publishing

Scholarship Name: ____________________

Contact Person: ____________________

Deadline: ____________________

Page No./Index No.: __________ / __________

Request Letter: __________, __________

Application Received: ____________________

Application Completed/Mailed:

__________ / __________

- __ Application Complete
- __ Official Transcript
- __ Letter(s) of Recommendation
- __ Essay(s)
- __ Reference Sheet
- __ Parents Tax Returns
- __ Return Envelope
- __ Application Signed & Dated
- __ Make Copies
- __ Other: ____________________

Special Notes: ____________________

Results: ____________________

Scholarship Track Form™
Flight Time Publishing

Scholarship Name: ____________________

Contact Person: ____________________

Deadline: ____________________

Page No./Index No.: __________ / __________

Request Letter: __________, __________

Application Received: ____________________

Application Completed/Mailed:

__________ / __________

- __ Application Complete
- __ Official Transcript
- __ Letter(s) of Recommendation
- __ Essay(s)
- __ Reference Sheet
- __ Parents Tax Returns
- __ Return Envelope
- __ Application Signed & Dated
- __ Make Copies
- __ Other: ____________________

Special Notes: ____________________

Results: ____________________

Scholarship Track Form™
Flight Time Publishing

Scholarship Name:________________________________

Contact Person: ________________________________

Deadline: ________________________________

Page No./Index No.: ______________ / ____________

Request Letter: ______________, ______________

Application Received: ____________________________

Application Completed/Mailed:

____________________ / ____________________

__ Application Complete
__ Official Transcript
__ Letter(s) of Recommendation
__ Essay(s)
__ Reference Sheet
__ Parents Tax Returns
__ Return Envelope
__ Application Signed & Dated
__ Make Copies
__ Other: ________________________________

Special Notes: ________________________________

Results:________________________________

Scholarship Track Form™
Flight Time Publishing

Scholarship Name:________________________________

Contact Person: ________________________________

Deadline: ________________________________

Page No./Index No.: ______________ / ____________

Request Letter: ______________, ______________

Application Received: ____________________________

Application Completed/Mailed:

____________________ / ____________________

__ Application Complete
__ Official Transcript
__ Letter(s) of Recommendation
__ Essay(s)
__ Reference Sheet
__ Parents Tax Returns
__ Return Envelope
__ Application Signed & Dated
__ Make Copies
__ Other: ________________________________

Special Notes: ________________________________

Results:________________________________

Scholarship Track Form™
Flight Time Publishing

Scholarship Name:________________________________

Contact Person: ________________________________

Deadline: ________________________________

Page No./Index No.: ______________ / ____________

Request Letter: ______________, ______________

Application Received: ____________________________

Application Completed/Mailed:

____________________ / ____________________

__ Application Complete
__ Official Transcript
__ Letter(s) of Recommendation
__ Essay(s)
__ Reference Sheet
__ Parents Tax Returns
__ Return Envelope
__ Application Signed & Dated
__ Make Copies
__ Other: ________________________________

Special Notes: ________________________________

Results:________________________________

Scholarship Track Form™
Flight Time Publishing

Scholarship Name: ____________________

Contact Person: ____________________

Deadline: ____________________

Page No./Index No.: __________ / __________

Request Letter: __________, __________

Application Received: ____________________

Application Completed/Mailed:

__________ / __________

- __ Application Complete
- __ Official Transcript
- __ Letter(s) of Recommendation
- __ Essay(s)
- __ Reference Sheet
- __ Parents Tax Returns
- __ Return Envelope
- __ Application Signed & Dated
- __ Make Copies
- __ Other: ____________________

Special Notes: ____________________

Results: ____________________

Scholarship Track Form™
Flight Time Publishing

Scholarship Name: ____________________

Contact Person: ____________________

Deadline: ____________________

Page No./Index No.: __________ / __________

Request Letter: __________, __________

Application Received: ____________________

Application Completed/Mailed:

__________ / __________

- __ Application Complete
- __ Official Transcript
- __ Letter(s) of Recommendation
- __ Essay(s)
- __ Reference Sheet
- __ Parents Tax Returns
- __ Return Envelope
- __ Application Signed & Dated
- __ Make Copies
- __ Other: ____________________

Special Notes: ____________________

Results: ____________________

Scholarship Track Form™
Flight Time Publishing

Scholarship Name: ____________________

Contact Person: ____________________

Deadline: ____________________

Page No./Index No.: __________ / __________

Request Letter: __________, __________

Application Received: ____________________

Application Completed/Mailed:

__________ / __________

- __ Application Complete
- __ Official Transcript
- __ Letter(s) of Recommendation
- __ Essay(s)
- __ Reference Sheet
- __ Parents Tax Returns
- __ Return Envelope
- __ Application Signed & Dated
- __ Make Copies
- __ Other: ____________________

Special Notes: ____________________

Results: ____________________

Scholarship Track Form™
Flight Time Publishing

Scholarship Name:______________________

Contact Person: ______________________

Deadline: ______________________

Page No./Index No.: __________ / __________

Request Letter: __________, __________

Application Received: ______________________

Application Completed/Mailed:

______________ / ______________

__ Application Complete
__ Official Transcript
__ Letter(s) of Recommendation
__ Essay(s)
__ Reference Sheet
__ Parents Tax Returns
__ Return Envelope
__ Application Signed & Dated
__ Make Copies
__ Other: ______________________

Special Notes: ______________________

Results:______________________

Scholarship Track Form™
Flight Time Publishing

Scholarship Name:______________________

Contact Person: ______________________

Deadline: ______________________

Page No./Index No.: __________ / __________

Request Letter: __________, __________

Application Received: ______________________

Application Completed/Mailed:

______________ / ______________

__ Application Complete
__ Official Transcript
__ Letter(s) of Recommendation
__ Essay(s)
__ Reference Sheet
__ Parents Tax Returns
__ Return Envelope
__ Application Signed & Dated
__ Make Copies
__ Other: ______________________

Special Notes: ______________________

Results:______________________

Scholarship Track Form™
Flight Time Publishing

Scholarship Name:______________________

Contact Person: ______________________

Deadline: ______________________

Page No./Index No.: __________ / __________

Request Letter: __________, __________

Application Received: ______________________

Application Completed/Mailed:

______________ / ______________

__ Application Complete
__ Official Transcript
__ Letter(s) of Recommendation
__ Essay(s)
__ Reference Sheet
__ Parents Tax Returns
__ Return Envelope
__ Application Signed & Dated
__ Make Copies
__ Other: ______________________

Special Notes: ______________________

Results:______________________

Scholarship Track Form™
Flight Time Publishing

Scholarship Name:______________________

Contact Person: ______________________

Deadline: ______________________

Page No./Index No.: __________ / __________

Request Letter: __________, __________

Application Received: ______________________

Application Completed/Mailed:

__________ / __________

__ Application Complete
__ Official Transcript
__ Letter(s) of Recommendation
__ Essay(s)
__ Reference Sheet
__ Parents Tax Returns
__ Return Envelope
__ Application Signed & Dated
__ Make Copies
__ Other: ______________________

Special Notes: ______________________

Results:______________________

Scholarship Track Form™
Flight Time Publishing

Scholarship Name:______________________

Contact Person: ______________________

Deadline: ______________________

Page No./Index No.: __________ / __________

Request Letter: __________, __________

Application Received: ______________________

Application Completed/Mailed:

__________ / __________

__ Application Complete
__ Official Transcript
__ Letter(s) of Recommendation
__ Essay(s)
__ Reference Sheet
__ Parents Tax Returns
__ Return Envelope
__ Application Signed & Dated
__ Make Copies
__ Other: ______________________

Special Notes: ______________________

Results:______________________

Scholarship Track Form™
Flight Time Publishing

Scholarship Name:______________________

Contact Person: ______________________

Deadline: ______________________

Page No./Index No.: __________ / __________

Request Letter: __________, __________

Application Received: ______________________

Application Completed/Mailed:

__________ / __________

__ Application Complete
__ Official Transcript
__ Letter(s) of Recommendation
__ Essay(s)
__ Reference Sheet
__ Parents Tax Returns
__ Return Envelope
__ Application Signed & Dated
__ Make Copies
__ Other: ______________________

Special Notes: ______________________

Results:______________________

Notes:

Notes:

Notes:

Notes:

Index

THIS PAGE INTENTIONALLY LEFT BLANK

A

B

C

D

E

F

N

O

P

Q

R

S

T

U

V

W

Additional Notes

Additional Notes

Additional Notes

Additional Notes

Additional Notes

Additional Notes

We want to hear from you!

Please tell us about any internships or scholarships programs you think we should know about. Then mail this form (feel free to attach additional sheets if necessary) to the addres listed below. Thanks!

Name of scholarship/internship: ______________________________

Contact person (if available): ______________________________

Address: ______________________________

____

Telephone: ____________________

Fax: ____________________

Website: ____________________

E-mail: ____________________

Mail To:

Flight Time Publishing
"The Source of Opportunities"™

8526 Drexel Ave., Suite 3B
Chicago, IL 60619

- -

We want to hear from you!

Please tell us about any internships or scholarships programs you think we should know about. Then mail this form (feel free to attach additional sheets if necessary) to the addres listed below. Thanks!

Name of scholarship/internship: ______________________________

Contact person (if available): ______________________________

Address: ______________________________

____

Telephone: ____________________

Fax: ____________________

Website: ____________________

E-mail: ____________________

Mail To:

Flight Time Publishing
"The Source of Opportunities"™

8526 Drexel Ave., Suite 3B
Chicago, IL 60619

We want to hear from you!

Please tell us about any internships or scholarships programs you think we should know about. Then mail this form (feel free to attach additional sheets if necessary) to the addres listed below. Thanks!

Name of scholarship/internship: ____________________

Contact person (if available): ____________________

Address: ____________________

____

Telephone: ____________________

Fax: ____________________

Website: ____________________

E-mail: ____________________

Mail To:

Flight Time Publishing
"The Source of Opportunities"™

8526 Drexel Ave., Suite 3B
Chicago, IL 60619

- - - - - - - - - - - - - - - - - - - -

We want to hear from you!

Please tell us about any internships or scholarships programs you think we should know about. Then mail this form (feel free to attach additional sheets if necessary) to the addres listed below. Thanks!

Name of scholarship/internship: ____________________

Contact person (if available): ____________________

Address: ____________________

____

Telephone: ____________________

Fax: ____________________

Website: ____________________

E-mail: ____________________

Mail To:

Flight Time Publishing
"The Source of Opportunities"™

8526 Drexel Ave., Suite 3B
Chicago, IL 60619

Flight Time Publishing
"The Source of Opportunities"™

ORDER FORM

Order your own copy of ***AVIATION SCHOLARSHIPS***, or order one for a friend.

☎ **Telephone Orders:** Call Toll Free 24 hours: 1 (800) 243-1515 ext. FTP (387). Have your VISA, MasterCard, or American Express card ready. **$2 OFF** when Code 4B is given.

✉ **Postal Orders:** Mail to: Flight Time Publishing, 8526 Drexel Ave., Suite 4B, Chicago, IL 60619-6210 USA

ALL POSTAL ORDERS, PLEASE ALLOW 3-4 WEEKS FOR DELIVERY.

** Community and civic organizations interested in volume discounts should write Flight Time Publishing "Request Department" on your organization's letterhead.*

Call Now 1-800-243-1515 ext. 387 or Visit Our Website at http://idt.net/~sdh19

$2 OFF (with code 4B) **Special Offer!!**

Cover price **$26.95** Please send me ____ copy(s) of ***AVIATION SCHOLARSHIPS.*** I am sending a check or money order along with this coupon for $24.95 plus $5 for shipping and handling for each copy ordered directly from the publisher. Mail to: **Flight Time Publishing**, 8526 Drexel Ave, Suite 4B, Chicago, IL 60619-6210

Name:______________________________
Address:____________________________
City:____________________St.__________
Zip:__________ Tel.___________________

Please add 8.75% to books ordered by Illinois Addresses

Flight Time Publishing
"The Source of Opportunities"™

ORDER FORM

Order your own copy of ***AVIATION SCHOLARSHIPS***, or order one for a friend.

☎ **Telephone Orders:** Call Toll Free 24 hours: 1 (800) 243-1515 ext. FTP (387). Have your VISA, MasterCard, or American Express card ready. **$2 OFF** when Code 4B is given.

✉ **Postal Orders:** Mail to: Flight Time Publishing, 8526 Drexel Ave., Suite 4B, Chicago, IL 60619-6210 USA

ALL POSTAL ORDERS, PLEASE ALLOW 3-4 WEEKS FOR DELIVERY.

** Community and civic organizations interested in volume discounts should write Flight Time Publishing "Request Department" on your organization's letterhead.*

Call Now 1-800-243-1515 ext. 387
or Visit Our Website at
http://idt.net/~sdh19

$2 OFF (with code 4B) **Special Offer!!**

Cover price **$26.95** Please send me ____ copy(s) of ***AVIATION SCHOLARSHIPS.*** I am sending a check or money order along with this coupon for $24.95 plus $5 for shipping and handling for each copy ordered directly from the publisher. Mail to: **Flight Time Publishing**, 8526 Drexel Ave, Suite 4B, Chicago, IL 60619-6210

Name:______________________________
Address:____________________________
City:____________________St.__________
Zip:__________ Tel.__________________
Please add 8.75% to books ordered by Illinois Addresses

Flight Time Publishing
"The Source of Opportunities"™

ORDER FORM

Order your own copy of ***AVIATION SCHOLARSHIPS***, or order one for a friend.

☎ **Telephone Orders:** Call Toll Free 24 hours: 1 (800) 243-1515 ext. FTP (387). Have your VISA, MasterCard, or American Express card ready. **$2 OFF** when Code 4B is given.

✉ **Postal Orders:** Mail to: Flight Time Publishing, 8526 Drexel Ave., Suite 4B, Chicago, IL 60619-6210 USA

ALL POSTAL ORDERS, PLEASE ALLOW 3-4 WEEKS FOR DELIVERY.

** Community and civic organizations interested in volume discounts should write Flight Time Publishing "Request Department" on your organization's letterhead.*

Call Now 1-800-243-1515 ext. 387
or Visit Our Website at
http://idt.net/~sdh19

$2 OFF (with code 4B) **Special Offer!!**

Cover price **$26.95** Please send me ____ copy(s) of ***AVIATION SCHOLARSHIPS.*** I am sending a check or money order along with this coupon for $24.95 plus $5 for shipping and handling for each copy ordered directly from the publisher. Mail to: **Flight Time Publishing**, 8526 Drexel Ave, Suite 4B, Chicago, IL 60619-6210

Name:______________________________
Address:____________________________
City:____________________St.__________
Zip:__________ Tel.__________________
Please add 8.75% to books ordered by Illinois Addresses